Transition Metal
Carbides and Nitrides

REFRACTORY MATERIALS

A SERIES OF MONOGRAPHS

John L. Margrave, *Editor*

DEPARTMENT OF CHEMISTRY
RICE UNIVERSITY, HOUSTON, TEXAS

Transition Metal Carbides and Nitrides

Louis E. Toth

CHEMICAL ENGINEERING—MATERIALS SCIENCE
UNIVERSITY OF MINNESOTA
MINNEAPOLIS, MINNESOTA

ACADEMIC PRESS   New York and London   1971

ACADEMIC PRESS, INC.
111 Fifth Avenue, New York, New York 10003

United Kingdom Edition published by
ACADEMIC PRESS, INC. (LONDON) LTD.
Berkeley Square House, London W1X 6BA

LIBRARY OF CONGRESS CATALOG CARD NUMBER: 72-127706

PRINTED IN THE UNITED STATES OF AMERICA

To Anna and Louis

Contents

4. Thermodynamics of Refractory Carbides and Nitrides

5. Mechanical Properties

6. Electrical and Magnetic Properties

7. Superconducting Properties

8. Band Structure and Bonding in Carbides and Nitrides

9.   Postscript

Preface

During the past five to ten years, the interest in transition-metal carbides and nitrides has increased substantially. Major advances have been made in our understanding of these compounds, not only in the more familiar areas of research such as crystallography, but also in areas in which there had previously been little investigation. As a result, knowledge about transition metal carbides and nitrides is extensive, but since this knowledge is reported in a great variety of journals, government reports, and conference proceedings, it is difficult to take full advantage of all the available information. Often a research investigator is not aware of some of the significant developments concerning these compounds in areas other than his specialty, or if he is aware of new developments, he may find it hard to place them in their proper perspective. This monograph attempts to present the recent developments in transition metal carbide and nitride research in a coherent fashion, and in some cases to evaluate it critically.

Selecting the topics to be considered and determining the amount of space to be devoted to each area has not been an easy task. The extensive literature on the technological uses of these materials has not been included; rather, emphasis has been placed on the more scientific areas. Within these areas there has been an attempt to balance the topics so that the more firmly developed research receives an in-depth treatment, while the more speculative results are considered only briefly or are not included at all. The areas of thermodynamic and mechanical properties are treated here, not with the intent of superseding the more exhaustive texts and reviews which already exist on these subjects, but to suggest possible new approaches to research in these fields. For many of the areas included, however, there has been no comprehensive review before now. Thus, the author has devoted a considerable part of this book to a survey of these previously unreviewed areas.

While this text should be most useful to research workers in metallurgy, ceramics, physics, chemistry, and related fields, advanced students investigating problems concerning high temperature materials or interstitial compounds might also benefit from this systematic presentation of information. The book grew out of a course on high temperature materials given by the author at the University of Minnesota. A course on very high temperature materials would, of course, include a treatment of oxides, borides, and superalloys, as well as carbides and nitrides; but, while adequate texts covering these other areas exist, previously no specialized text has been available in the area of carbides and nitrides.

Acknowledgments

A number of people reviewed parts of the text and offered many helpful suggestions. At the University of Minnesota, Professor T. Zoltai reviewed the section on crystal structure, and Professors D. F. Stein and M. E. Nicholson reviewed the chapter on mechanical properties. This chapter was also reviewed by Professor W. S. Williams at the University of Illinois and by Dr. G. Hollox at Brown Boveri & Cie, Baden, Switzerland. Professor Y. A. Chang at the University of Wisconsin, Milwaukee, reviewed the chapter on thermodynamic properties.

The author is also indebted to Dr. E. K. Storms of the Los Alamos Scientific Laboratory for reviewing the entire manuscript, for offering many helpful suggestions, and for allowing the publication here of many of his thermodynamic evaluations.

My wife Susan kindly helped on many questions about writing style. Mrs. Christina P. Meyer, Mrs. Marie Y. Wolf, and Miss Susan Kirkpatrick, graduate students in English at the University of Minnesota, helped considerably in the preparation of the book, and Mrs. Dorothy A. Trapp did much of the typing.

Transition Metal Carbides and Nitrides

I

General Properties, Preparation, and Characterization

I. Introduction

The transition-metal carbides and nitrides discussed in this text are those of the fourth to sixth group of the periodic table. Most of the carbides and nitrides in this group have extremely high melting points (2000–4000°C) and therefore these compounds are frequently referred to as "refractory carbides and nitrides." Several of the sixth group nitrides are exceptions in that they decompose at relatively low temperatures. While these materials are refractory, presently, their main commercial importance stems from their extreme hardness. The carbides in this group form the basis for "cemented carbide" cutting tools and wear-resistant parts. Because they have excellent high-temperature strength and good corrosion resistance, they can also be used as high-temperature structural materials. The nitrides of this group are useful because of their hardness, but they are being used increasingly for their electrical properties in applications such as integrated circuitry. These nitrides are also superconductors with attractive properties for potential applications.

Carbide and nitride formation is fairly common among transition metals, as illustrated in Tables I and II. In the first transition series, every element forms at least one carbide and nitride; in the second and third transition series, carbide and nitride formation is restricted mainly to elements from

1

TABLE I

CARBIDE FORMATION IS FAIRLY COMMON AMONG THE TRANSITION ELEMENTS, EXCEPT FOR THE SECOND AND THIRD ROWS OF GROUP VIII[a]

III	IV	V	VI	VII	VIII		
$Sc_{2-3}C$ ScC_2 Sc_2C_3	TiC	V_2C VC	$Cr_{23}C_6$ Cr_7C_3 Cr_3C_2	$Mn_{23}C_6$ Mn_3C Mn_5C_2 Mn_7C_3	Fe_3C	Co_3C Co_2C	Ni_3C
Y_2C Y_2C_3 YC_2	ZrC	Nb_2C NbC	Mo_2C Mo_3C_2 MoC_{1-x}	TcC	Ru ⊠	Rh ⊠	Pd ⊠
LaC_2	HfC	Ta_2C TaC	W_2C W_3C_2 WC	ReC	OsC	Ir ⊠	Pt ⊠

[a] ⊠ indicates no carbide formation for this element.

groups three to seven. Carbides and nitrides of Mn, Fe, Co, and Ni are important constituents in steels; they are the hard second phases that strengthen the steels. These carbides and nitrides are not included in this text because

TABLE II

NITRIDE FORMATION IS ALSO FAIRLY COMMON AMONG THE TRANSITION ELEMENTS, EXCEPT FOR THE SECOND AND THIRD ROWS OF GROUP VIII[a]

III	IV	V	VI	VII	VIII		
ScN	Ti_2N TiN	V_2N VN	Cr_2N CrN	Mn_4N Mn_2N Mn_3N_2	Fe_4N Fe_2N	Co_3N Co_2N Co_3N_2	Ni_3N Ni_3N_2
YN	ZrN	Nb_2N Nb_4N_3 NbN_{1-x} NbN	Mo_2N MoN	TcN	Ru ⊠	Rh ⊠	Pd ⊠
LaN	HfN	Ta_2N TaN Ta_3N_5	W_2N WN	Re_2N	Os ⊠	Ir ⊠	Pt ⊠

[a] ⊠ indicates no nitride formation for this element.

a discussion of their properties cannot be isolated from a discussion of steels. Such a task would be well beyond the scope of this book.

Discussing the properties of carbides and nitrides together in a text of this type is justified on the basis of a similarity in structure and properties.

Of all compounds formed between transition metal atoms and light elements H, B, C, N, and O, only carbides and nitrides are closely related in crystal structure types, phase relationships, bonding characteristics, and electric and magnetic properties. This close relationship between transition-metal carbides and nitrides is easy to understand owing to similarities of electron structure, size, and electronegativity of carbon and nitrogen atoms. These similarities do not extend to transition-metal borides, which are characteristically distinguished by boron–boron atomic bonds, as indicated by the proximity of the atoms in the crystal structures. Nor do these similarities extend to more than a few transition-metal oxides, which are characteristically ionic in bonding, structure, and properties.

Despite the convenience of discussing carbides and nitrides together, it is important to remember that the basic difference between carbon and nitrogen, one additional electron, is also reflected in the transition metal compounds. Because of this additional electron, the general properties of fifth-group nitrides are often more similar to those of sixth-group carbides than to fifth-group carbides. Discussion of these similarities and differences between carbides and nitrides forms one basis of this text.

The focus of this monograph is considerably different from those of previous texts on the same materials (1–10). Some texts have emphasized the crystal chemistry of carbides and nitrides, some have emphasized their thermodynamic properties, and others have emphasized their applications. This monograph, however, relates structure (crystal, defect, and electronic) to properties (thermodynamic, mechanical, electrical, magnetic, and superconducting). In terms of these structures, such problems as why carbides and nitrides are refractory, why they have great hardness and strength, and why they share certain electrical and superconducting properties are treated. Many recent technological advances in the development of these materials for unique mechanical and electrical properties can be coupled directly to an increased understanding of the interatomic bonding and defect structure.

One of the significant developments in the understanding and expanding uses of these materials is an increased awareness of the importance of defect structure on the properties. The carbides and nitrides are not primarily stoichiometric phases. The composition range for each binary phase is extensive, and all types of properties depend upon the nonmetal-to-metal ratio and the vacancy concentration. Failure to realize these dependencies has resulted in many erroneous literature reports and much confusion about certain properties. Not only are the compositional dependencies emphasized in this text, but also the newly discovered ordering possibilities of carbon and nitrogen in the defective phases, and the resultant property changes are discussed. The compound V_6C_5, for instance, undergoes a marked change in mechanical strength on ordering of the carbon atoms. Understanding this ordering is a recent development involving sophisticated characterization techniques of neutron diffraction, electron microscopy, and advanced X-ray diffraction.

Much of the discussion of the crystal chemistry of carbides and nitrides concerns the nonstoichiometric phases and the possibilities for ordering whenever a sizable fraction of interstitial sites are vacant. The crystal chemistry is also treated from the viewpoint of a few coordination polyhedra. Not only can most carbide and nitride crystal structures be reconstructed from these polyhedra, but their type and geometric arrangement affect the atomic bonding and, hence, their properties.

The atomic bonding is treated from the viewpoint of the electronic band structure. This approach, while more complex than previous treatments in other texts, is necessary if one is to gain more than a superficial understanding. While the proposed band structures are still controversial, they do allow experimental verification; thus, by correlating several types of properties with the various band models, one can develop a fair concept of bonding in these compounds. With this increased knowledge, better materials with particular technological useful properties have been developed. In the future, this approach should result in a rapid development in our understanding of these materials and in their expanding uses.

II. Survey of General Properties

Their unusual properties make transition-metal carbides and nitrides both interesting and useful. Several general properties are described below; these properties will be discussed in more detail in the subsequent chapters.

1. An important property of carbides and nitrides is their great hardness, for these compounds are among the hardest known. Many binary carbides have microhardness values between 2000 and 3000 kg/mm²—values which lie between those of Al_2O_3 and diamond. This property has resulted in the extensive use of carbides as cutting tools and for wear-resistant surfaces. The nitrides, while hard, are not as hard as the carbides and are not used extensively for this property. Table III (p. 6) lists some typical microhardness values for binary carbides and nitrides as well as other properties (*11*).

2. A second striking property of transition-metal carbides and nitrides is their very high melting points. Several carbides and nitrides melt or decompose above 3000°C and TaC has the highest melting point known for any material, about 3983°C (graphite sublimes at about 4000°C). The melting points of the carbides are generally higher than those of the parent transition-metal elements; the melting or decomposition temperatures of the nitrides are comparable to those of the pure transition-metal elements. Figure 1 compares the melting point of carbides and nitrides with those of transition metals of the same groups. There is an interesting displacement in group

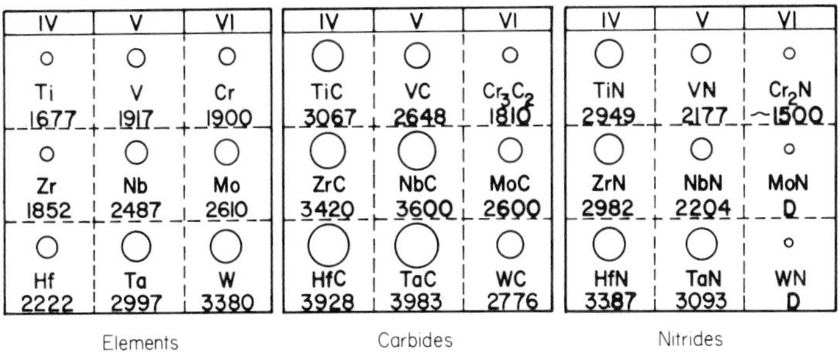

IV	V	VI		IV	V	VI		IV	V	VI
○	○	○		○	○	○		○	○	○
Ti	V	Cr		TiC	VC	Cr_3C_2		TiN	VN	Cr_2N
1677	1917	1900		3067	2648	1810		2949	2177	~1500
○	○	○		○	○	○		○	○	○
Zr	Nb	Mo		ZrC	NbC	MoC		ZrN	NbN	MoN
1852	2487	2610		3420	3600	2600		2982	2204	D
○	○	○		○	○	○		○	○	○
Hf	Ta	W		HfC	TaC	WC		HfN	TaN	WN
2222	2997	3380		3928	3983	2776		3387	3093	D
	Elements				Carbides				Nitrides	

Fig. 1. Comparison of melting points for elements, carbides, and nitrides. The size of the circle is proportional to the melting point. The maximum melting points occur at group VI for the elements, at group V for carbides, and at group IV for nitrides. The letter "D" means that the material decomposes at a relatively low temperature, less than 800°C.

number as to where the maxima in melting points occur. In the transition metals, group VI elements have the highest melting point; for the carbides the maxima lie at group V; and for the nitrides, group IV has the highest decomposition temperatures under 1 atm of nitrogen pressure.

3. Perhaps the most important property of this group of carbides and nitrides is their defect structure. Ideal stoichiometry is generally not found in these phases; deviations from stoichiometry are far more common. The phases exist over broad composition ranges and appreciable vacancy concentrations (up to 50 at. %) can exist on the nonmetal lattice sites with lesser concentrations on metal-atom lattice sites. Even at the stoichiometric ratio appreciable vacancy concentrations (a few atomic percent) can exist on both sublattices in certain cases. When a large fraction of nonmetal sites are vacant, the vacancies tend to exhibit long-range order. This phenomenon, only recently appreciated, is found in nearly all the subcarbides, subnitrides, and in several monocarbides. The presence of a large concentration of vacancies, ordered or disordered, significantly affects properties—thermodynamic, mechanical, electrical, magnetic, superconducting, etc. Different processing techniques tend to produce different defect structures and hence different properties.

TABLE III
SURVEY OF SELECTED PROPERTIES OF TRANSITION-METAL CARBIDES AND NITRIDES[a,b]

Phase	Structure	Lattice parameters, Å	X-ray density, gm/cm³	Micro-hardness kg/mm²	Melting or decomposition temp., °C	Heat conductivity, cal/cm-sec, °C	Coefficient of thermal expansion $\times 10^{-6}$, room temp.	Color
TiC	B1	4.328	4.91	2900	3067	0.05	7.4	Gray
ZrC$_{0.97}$	B1	4.698	6.59	2600	3420	—	6.7	Gray
HfC$_{0.99}$	B1	4.640	12.67	2700	3928	—	6.6	Gray
VC$_{0.87}$	B1	4.166	5.65	2900	2648	—	—	Gray
NbC$_{0.99}$	B1	4.470	7.79	2400	3600	0.034	6.6	Lavender
TaC$_{0.99}$	B1	4.456	14.5	2500	3983	0.053	6.3	Golden
Cr$_3$C$_2$	Ortho	a: 11.47 b: 5.545 c: 2.830	6.68	1300	1810	—	10.3	Gray
Mo$_2$C	Ortho	a: 7.244 b: 6.004 c: 5.199	9.06	1500	—	—	4.9//a 8.2//c	Gray

WC	Hex	a: 2.906 c: 2.837	15.8	2100 on basal plane	2776	0.07	5.0//a 4.2//c	Gray
TiN	$B1$	4.240	5.39	2000	2949	0.046	9.35	Golden-yellow
ZrN	$B1$	4.577	7.3	1500	2982	0.049	7.24	Golden
HfN	$B1$	4.526	13.8	1600	3387	0.052	6.9	Golden-green
VN	$B1$	4.14	6.0	1500	2177	0.027	8.1	Gray-red-brown
NbN	Hex	a: 2.958 c: 11.272	7.3	1400	2204	0.009	10.1	Gray
TaN	Hex	a: 5.185 c: 2.908	14.3	1000	3093	0.021	3.6	Gray
CrN	$B1$	4.149	6.1	1100	1500	0.028	2.3	Gray-brown

[a] See references 1, 4, 5, and 11.

[b] Lattice parameters and density refer to compositions closest to stoichiometry.

4. Because of possibilities of vacancies ordering in the defect structures, the crystal chemistry of carbides and nitrides is complex. Carbon and nitrogen, being smaller atoms than the transition-metal elements, are always interstitially located. One may view the crystal structures as the geometric arrangement of interstitial atoms and vacancies in a relatively simple metal-atom structure or alternatively as a geometric arrangement of coordination polyhedra, the principal one being an octahedral grouping of metal atoms around a carbon or nitrogen atom. The former approach is ideally suited for describing the vacancy defect structures while the latter approach is useful in comparing simple structural similarities between otherwise complex structures. While many of the crystal structures of carbides and nitrides are complex, most of them can be reconstructed from only a few different coordination polyhedra.

5. They are chemically stable at room temperature and are attacked slowly only by very concentrated acid solutions. An exception to this general statement is VC, which is slowly attacked by air at room temperature. At high temperatures they oxidize readily to form oxides. Both their chemical reactivity and thermodynamical properties are sensitive functions of the nonmetal-to-metal ratio. Many of these dependencies have still not been investigated. Because of the complex nature of the defect structure, a theoretical treatment of their thermodynamical properties is difficult and to date, relatively unsuccessful.

6. The carbides of this group are extremely strong particularly at high temperatures. They possess high values of Young's modulus, $40-90 \times 10^6$ psi, compared to $20-40 \times 10^6$ psi for most transition-metal elements. At room temperature they are brittle, but at high temperatures (about 1000°C) they undergo a brittle-to-ductile transition, and above that temperature they plastically deform in a manner analogous to that for fcc metals. This feature is important because it means that polycrystalline carbides can also be made ductile. Above the brittle-to-ductile transition temperature the carbides are among the strongest materials known. At 1250°C V_6C_5 has a compressive yield strength-to-density ratio of 400,000 in. (*12*), and at 1800°C a two-phase alloy of VC_{1-x} and TiC has one of the highest compressive yield strengths of any known material, 36,000 psi (*13*).

7. Carbides and nitrides of this group are typically metallic in their electrical, magnetic, and optical properties. Most of these properties differ only slightly from those of the parent transition-metal elements. The electrical and magnetic properties of carbides and nitrides are extremely sensitive to defect structure, principally vacancies on both nonmetal and metal lattice sites. Probably because of large vacancy concentrations, the temperature dependence of the electrical and thermal conductivities of carbides and nitrides differ considerably from those of the parent transition-metal elements. The electrical resistivities of carbides and nitrides show little, if any, temperature dependence, a feature which has current applications.

8. The nitrides, in particular, are an interesting class of superconductors. NbN based alloys have some of the highest superconducting critical temperatures (T_c), about 18°K—the highest known T_c is currently 20°K. While superconductivity is fairly common among elements, compounds, and alloys, T_c's above 10°K are relatively rare. The occurrence of superconductivity with a high T_c is very common among carbides and nitrides. The NbN alloys also have excellent upper critical fields and critical currents. Superconductivity exists in these alloys in magnetic fields of over 200 kG and current densities of 10^5 A/cm² are observed even in fields of 100 kG. The superconducting parameters are sensitive to nonmetal-to-metal ratio, defect structure, and methods of preparation. In many cases the relationship between composition and defect parameters with a particular superconducting parameter has not been firmly established. While the nitrides have excellent superconducting properties, they have not been used extensively in superconducting devices. Thin-film nitrides seem to have the most potential for application in such devices as Josephson junctions.

9. Bonding in these phases consists of a complex combination of localized metal-to-metal and metal-to-nonmetal interactions resembling both covalent and metallic bonding. There is only a small amount of ionic bonding with the amount being slightly greater in the nitrides than in the carbides. The metal-to-nonmetal bonding is favored by the octahedral grouping of metal atoms around a central carbon or nitrogen atom, but the presence of the nonmetal in this grouping also tends to increase the strength of metal-to-metal bonds. The bonding mechanism in these phases is controversial although recent band structure calculations do agree on many aspects. The most important problem, the variation in bonding with defect structure, has not been attempted; a similar problem in TiO has been calculated and the solution indicates major changes in the bonding with changes in the defect structure.

III. Applications

The principal application for carbides is as the major constituent in "cemented carbide" cutting tools. The term "cemented carbide" refers to a fourth to sixth group carbide bonded together in a metal matrix, usually cobalt. WC is the most important carbide for this application, over 1,000,000 lb being used annually (*11*). Other carbides used for this purpose include TiC, TaC, and NbC usually as alloying additions to WC. Cobalt in the amounts 5–30 wt % is added to increase the toughness of the tool bit without greatly reducing the hardness and it is also added because of the inability of carbides to self-sinter except under the application of very high temperatures.

Carbides have a unique set of properties necessary for use as a cutting tool. Property requirements include great hardness and wear resistance, good thermal shock resistance and thermal conductivity, good oxidation resistance, and compatibility of the carbide particles with the cobalt binder. Good thermal shock resistance and thermal conductivity are necessary to remove local heat generated at cutting surfaces. Since carbides are metallic conductors, this property is assured because of the rapid heat conduction by the free electrons. Nonmetallic ceramics generally have poorer thermal conductivity and shock resistance. Compatability of carbides with the binder materials, usually cobalt, is also excellent. Cobalt has the ability of sufficiently wetting the surface of carbide particles to ensure good binding, but the solubility of cobalt in the carbide is low enough so that the carbide is left relatively pure after sintering.

Additions of TiC, TaC, and NbC cause important changes in the properties of cemented carbides. Pure WC–Co cemented carbides wear rapidly because of a local welding of the tool with the steel part to be cut. Addition of TiC causes the formation of a TiO surface layer that effectively isolates the tool from the part to be cut and, in a sense, protects the tool from rapid wear. The compound WC forms an oxide WO_3, but it is volatile and therefore offers no protection. Additions of TaC and NbC raise the melting temperature of the carbide solution and increase the oxidation resistance over that of (WC, TiC)–Co.

Besides their use as cutting tools, the cemented carbides are used extensively as spikes for snow tires, as wear resistant parts in wire drawing, extrusion and pressing dies, as drilling tools in mining, and as wear resistant surfaces in many types of machines and instruments.

In high-temperature applications, carbides are used either as self-bonded sintered parts or as a constituent in a sintered composite consisting of carbide particles bonded together with Co, Mo, or W. Typical parts include rocket nozzles and jet engine components. These materials are used where normal superalloys are no longer applicable because the required operating temperatures are too high. Self-bonded carbides such as TiC_{1-x} and VC_{1-x} alloys have great strengths up to about 1800°C and can be used as high temperature structural materials. Below about 1000°C these carbides are brittle and surface imperfections or internal pores can greatly reduce strength. Care must be taken to remove these surface flaws and to use fully dense sintered carbides.

In refractory alloys of Nb, Ta, Cr, Mo, and W, carbides are used as a minor constituent to act as dispersion strengtheners. With increasing temperature, the solubility of carbon increases in these metals and therefore precipitation hardening is possible by controlled heat treatments.

Despite their excellent high temperature properties, good corrosion resistance, and hardness, the nitrides of this group are not used so extensively as the carbides. Wear resistant surfaces of TiN can be imparted to Ti parts by nitriding. With the increasing use of Ti as a structural material, this use of the nitride should increase.

The major future use of the nitrides may depend upon their electrical properties. In integrated circuits nitride thin-films are easily deposited by reactive sputtering. The electrical resistivity of films of TaN and NbN based alloys have nearly zero temperature dependence and this property coupled with their excellent corrosion resistance should result in expanded use of nitrides.

Another potential use for nitrides is in superconducting devices. Nitrides have been suggested as materials for thin-film miniature superconducting solenoids, high-Q inductors, Josephson junctions, and bolometers. These materials are particularly promising for Josephson junctions; their refractory nature and corrosion resistance results in little chemical diffusion and hence decay of the junction with time. Junctions in use today are not chemically stable after cycling a few times between room and liquid helium temperatures.

IV. Preparation Techniques

Polycrystalline samples of carbides and nitrides are usually prepared by powder metallurgy. The metal or oxide powder is reacted with carbon or nitrogen, pressed, and sintered. Since these techniques are extensively covered in other texts (1, 2, 4), we will only outline some of the general powder metallurgy techniques and concentrate instead on the rapidly growing techniques for single crystals and thin films.

IVA. Powder Metallurgy Techniques

Tables IV and V list the more common powder-metallurgy techniques for preparing polycrystalline carbides and nitrides. Direct reaction of metal or metal hydride powders with carbon or nitrogen is often the most common laboratory method for the formation of nearly all carbides and nitrides. This technique is also one of the best for producing relatively pure compounds. The preparation of high-purity homogeneous samples for research purposes is, however, a difficult task. Very high temperatures and good vacuum conditions or highly purified gases are generally required. The preparation conditions, furthermore, vary according to the alloy system and

TABLE IV

PREPARATION OF METAL CARBIDES[a]

Method	Reaction
(1) Direct reaction either by melting or sintering of the elements or metal hydrides in a protected atmosphere or in vacuum.	$Me + C \rightarrow MeC$ $MeH + C \rightarrow MeC + H_2$
(2) Direct reaction of the metal oxide and excess carbon in a protective or reducing atmosphere.	$MeO + C \rightarrow MeC + CO$
(3) Reaction of the metal with a carburizing gas.	$Me + C_x H_y \rightarrow MeC + H_2$ $Me + CO \rightarrow MeC + CO_2$
(4) Precipitation from the gas phase by reacting the metal halide or metal carbonyl in hydrogen.	$MeCl_4 + C_x H_y + H_2$ $\rightarrow MeC + HCl + (C_m H_n)$ $Me\text{--}Carbonyl + H_2 \rightarrow MeC$ $+ (CO, CO_2, H_2, H_2O)$

[a] See Kieffer and Schwarzkopf (1).

TABLE V

PREPARATION OF METAL NITRIDES[a]

Method	Reaction
(1) Nitriding the metal powders or metal hydride powders.	$Me + N_2 \rightarrow MeN$ $MeH + N_2 \rightarrow MeN + H_2$ $Me(MeH) + NH_3 \rightarrow MeN + H_2$
(2) Nitriding metal oxide powders in the presence of carbon.	$MeO + N_2(NH_3) + C \rightarrow MeN + CO$ $+ H_2O + H_2$
(3) Reaction of metal chlorides and NH_3.	$MeCl_4 + NH_3 \rightarrow MeN + HCl$ $MeOCl_3 + NH_3 \rightarrow MeN + HCl$ $+ H_2O + H_2$
(4) Precipitation from the gas phase by reacting the metal halide in a $N_2\text{--}H_2$ atmosphere.	$MeCl_4 + N_2 + H_2 \rightarrow MeN + HCl$

[a] See Kieffer and Schwarzkopf (1).

what may be an acceptable procedure in one case may result in an impure or inhomogeneous sample in another. The techniques for the carbides have been comprehensively reviewed by Storms (4) and the interested reader is referred to his book *The Refractory Carbides* for an up-to-date description of the appropriate technique to utilize for a given alloy. Other readers interested primarily in the technological uses of carbides as cutting tool components are referred to the preparation techniques described by Kieffer *et al.* (1) and Kieffer and Benesovsky (11).

Typical temperatures for reacting these powders are listed in Tables VI and VII. These temperatures are not the lowest at which some reaction will

TABLE VI

REACTION TEMPERATURES FOR THE FORMATION OF
REFRACTORY CARBIDES FROM THE ELEMENTS[a]

Reaction	Reaction temperature (°C)
$Ti(TiH_2) + C \rightarrow TiC$	1700–2100
$Zr(ZrH_2) + C \rightarrow ZrC$	1800–2200
$Hf(HfH_2) + C \rightarrow HfC$	1900–2300
$V + C \rightarrow VC$	1100–1200
$Nb + C \rightarrow NbC$	1300–1400
$Ta + C \rightarrow TaC$	1300–1500
$Cr + C \rightarrow Cr_3C_2$	1400–1800
$Mo + C \rightarrow MoC$	1200–1400
$W + C \rightarrow WC$	1400–1600

[a] See Schwarzkopf and Kieffer (2).

TABLE VII

REACTION TEMPERATURES FOR THE FORMATION OF
REFRACTORY NITRIDES[a]

Reaction	Reaction temperature (°C)
$Ti(TiH_2) + N_2 \rightarrow TiN$	1200
$Zr(ZrH_2) + N_2 \rightarrow ZrN$	1200
$Hf(HfH_2) + N_2 \rightarrow HfN$	1200
$V + N_2 \rightarrow VN$	1200
$Nb + N_2 \rightarrow NbN$	1200
$Ta + N_2 \rightarrow TaN$	1100–1200
$Cr + NH_3 \rightarrow CrN + H_2$	800–1000
$Mo + NH_3 \rightarrow MoN + H_2$	400–700
$W + NH_3 \rightarrow WN + H_2$	700–800

[a] See Schwarzkopf and Kieffer (2)

occur but rather those temperatures required for reaction within a relatively short period of time. Hafnium powder, for example, reacts with nitrogen at atmospheric pressure at 700°C to form nonhomogeneous $HfN_{0.4}$ within about 3 hr, but at 1200°C nearly stoichiometric HfN forms in the same period of time. Temperature and time are crucial variables in determining the final composition and homogeneity of the carbide and nitride. While the temperatures in Tables VI and VII will result in good yields of carbide and nitride formation, the listed temperatures are generally too low to prepare homogeneous samples. Most carbide systems, for instance, must be heated to greater than 2000°C for several hours to ensure a homogeneous product. Under certain conditions, such as a good vacuum, the carbide can also be purified from oxygen contamination by this high-temperature treatment. The homogenization and purification process varies according to the alloy system, and the reader should consult Storms' book for the particular process that is most desirable.

One common method for preparing homogeneous substoichiometric nitrides is to nitride the metal powders at a lower temperature than indicated in Table VII for a short period of time and then to homogenize the powders at an elevated temperature in an inert atmosphere. Homogenizing, however, is best done under an equilibrium nitrogen partial pressure to control the final nitrogen composition.

Substoichiometric monocarbides and subcarbides can be prepared by directly reacting metal powders and carbon mixed to the desired composition. Alternatively, the nearly stoichiometric monocarbide can be reacted with metal powders to produce the substoichiometric monocarbides and subcarbides.

Sixth-group elements cannot effectively be nitrided in N_2. They form nitrides at low temperatures only in an NH_3 atmosphere, and they require very long reaction times (2 or 3 weeks for fine powders). At temperatures higher than those listed in Table VII, the sixth-group nitrides decompose.

IVB. Preparation of Single Crystals

Recent interest in the mechanical and electrical properties of carbides and nitrides has necessitated the preparation of single crystals. Carbide single crystals have been prepared by both the Verneuil technique and the floating zone technique; they have also been precipitated from liquid metals. Nitride single crystals have been grown by vapor transport processes.

Single crystals of WC are readily grown from Co–WC molten mixtures (14–18). Takahashi and Freise (17) describe the following technique: Co–WC mixtures (40 wt % WC) are melted in alumina crucibles at 1600°C under argon. The temperature is held at 1600°C for 1–12 hr to ensure homogeneity, and is then lowered to 1500°C at the rate of 1–3°C/min. The temperature is maintained at 1500°C for an additional 12 hr before the sample is allowed to furnace-cool. The cobalt matrix is then dissolved away in boiling HCl. Both triangular and rectangular platelets are formed; the largest platelets are about 5 mm.

Fig. 2. Tungsten carbide crystal 1 cm on a side grown in the ⟨1100⟩ direction. [After Gerk and Gilman (*18*).]

Larger and better quality WC crystals have been grown using a modified Czochralski technique (*18*). A WC–Co mixture is homogenized well above the liquidus temperature. The liquid is then slowly cooled under a vertical temperature gradient until the colder top surface is near the liquidus temperature. A seed crystal is lowered to the surface and slowly withdrawn from the melt at a rate of about 10 mil/hr. As the crystal grows and the melt becomes depleted of WC, the surface temperature is changed to maintain the liquidus temperature. Crystals about 1 cm have been grown by this technique. These crystals produce sharp Laue patterns, indicating little residual strain, and they contain few Co inclusions (see Fig. 2).

Single crystals of Mo_2C, TiC, ZrC, and NbC have been successfully grown by the Verneuil technique (*19*). In the original Verneuil technique, powder is dropped into a stream of an oxygen–hydrogen burner. The powder melts as it emerges from the flame, and the liquid, deposited in a conical shape, tends to solidify as a single crystal. In preparing carbides, the combustion heat source is replaced by an electric arc, and the powers are protected against air with a stream of argon. This method is not ideal, since relatively small (millimeter size) single crystals are produced, and during cooling the boule tends to crack. Other problems include controlling the shift between starting and final compositions and controlling the size and composition of the starting powder.

Single crystals of VC_{1-x} and TiC_{1-x} have been grown by a conventional floating zone technique under 10 atm of helium (*20*). Isostatically pressed and sintered rods of VC_{1-x} are heated with an rf induction unit to produce a 1-cm-high molten zone. The zone is then moved along the rod at a rate of 1 cm/hr. An afterheater (graphite susceptor) reduces thermal shock. Crystals of about 1 cm in diameter and 8 cm in length can be produced in this manner (see Figs. 3 and 4).

Nearly stoichiometric single crystals of ZrC have been grown by direct reaction of the elements at very high temperatures (*21*). A Zr rod is inserted into a tight fitting graphite chamber and the chamber is then capped with a graphite plug (see Fig. 5). The graphite and rod are heated above the melting point of Zr. A carbide forms on the outer surface of the metal by direct chemical reaction and inside the rod by diffusion of carbon through the outer carbide layer. Many incremental anneals at increasing temperatures result in large grains or even crystals the size of the initial Zr rod. The mechanism for grain growth is a variation of the strain–anneal technique. The carbide is under considerable stress because of the difference in thermal expansion, induced by the increasing temperatures, between the growing carbide and the incompletely reacted liquid metal contained by the carbide. The stress plastically deforms the carbide. Prolonged heating at higher temperatures promotes recrystallization to reduce the strain energy. This

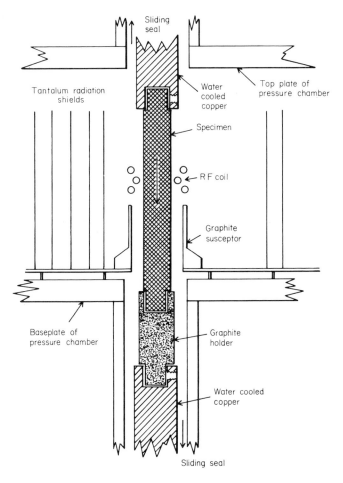

Fig. 3. Schematic representation of the zone refining apparatus used to grow single crystals of TiC and VC. [After Precht and Hollox (20).]

method was successfully applied to ZrC and VC, but it was not successful for HfC, TaC, and Ta$_2$C.

In a second variant of the strain–anneal technique, fully dense carbide rods are given a critical strain in axial compression at very high temperatures ($\sim$2500°C for HfC). Recrystallization occurs at higher annealing temperatures ($\sim$2900°C for HfC) to relieve the strain energy. A critical plastic strain results in very large grains or even single crystals $\frac{1}{4}$ in. in diameter and 1 in. long. The method was successfully applied to ZrC, TaC, HfC, and

Ta$_2$C (*21*). Many of the crystals grown by this technique are now commercially available.[1]

One major advantage of the strain–anneal techniques is the ability to prepare nearly stoichiometric single-crystal carbides. In the floating-zone technique, the crystals tend to be carbon deficient. In both techniques, good quality strain-free crystals are produced.

Very few references in the literature describe the preparation of nitride single crystals. Brager (*22*) reports that single crystals of TiN can be vapor-deposited on a copper plate from a gas phase consisting of NH$_3$ and TiCl$_4$ at 800°C. Single crystals of ZrN and TiN can be deposited on hot tungsten filaments by the decomposition of the tetrachlorides in the presence of N$_2$ and H$_2$ or NH$_3$ (*23*).

IVC. Thin Films

Very thin films of carbides and nitrides are useful in a number of applications, particularly as thin-film components (temperature insensitive precision resistors and capacitors) for the electronics industry. The films are also very useful for electron diffraction studies and crystal structure determinations. This technique for carbides and nitrides makes it possible to determine the atomic positions of both metal and nonmetal atoms in the crystal structure. In contrast, X-ray techniques usually determine only metal atoms, due to the large difference in atomic scattering parameter. Other applications for very thin films ($\sim$1000–10,000 Å) of carbides and nitrides include superconducting devices (see Chapter 7).

Thin films of Mo and W nitrides have been prepared by Troitskaya and Pinsker by first forming Mo or W films by sublimation and subsequently nitriding the films in NH$_3$ at about 750°C (*24, 25*). Unfortunately, details of this particular technique were not fully disclosed.

If relatively pure nitrides are to result from this method, it is necessary to take several precautions to avoid contamination of the film. Preferably, an electron beam should be used to prepare films of the pure transition-metal element. Since only a small part of the sample is molten at one time, the problem of finding a suitable crucible to hold the refractory element is

[1] These crystals are now available from Alpha Crystals Inc.

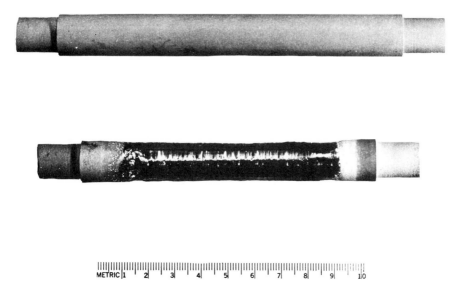

Fig. 4. A typical crystal of VC produced by conventional zone refining techniques. The upper rod is the sintered starting material; the lower rod is the resulting single crystal. [After Precht and Hollox (20).]

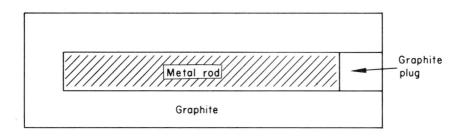

Fig. 5. To prepare single crystals of ZrC by direct reaction of the elements, a Zr rod is inserted into a tight-fitting graphite tube and cap. As the rod melts, the Zr forms a carbide that is under considerable stress. The carbide continues to grow by carbon diffusion. With incremental anneals at progressively higher temperatures, recrystallization occurs. Under proper conditions significant grain growth also occurs (21).

partially avoided. Then the evaporation unit should be processed as an ultrahigh vacuum unit. A vacuum of 10^{-6} Torr may not be sufficient to deposit a relatively pure film of the parent element, particularly if the element is from the fourth group. Third, the nitriding atmosphere should be of spectroscopic grade purity. Finally, contamination from the substrate may occur at elevated nitriding temperatures; and this problem may be difficult to overcome.

The high temperatures used to form nitrides in the above technique can be avoided by reactive sputtering (26–37). Reactive sputtering can be defined as sputtering in the presence of a reactive gas so that the deposited film has a different chemical composition than the target material. To form carbides and nitrides, the reactive gas is either a carbon containing gas or N_2.

Since all the metals of the fourth to sixth groups can be sputtered, an almost unlimited number of binary, ternary, or higher-order carbides and nitrides can be deposited by reactive sputtering. It is also possible to vary the nonmetal-to-metal ratio by varying the partial pressure of the reactive gas.

Most reactive sputtering studies on the refractory compounds have used the dc glow discharge technique (cathode sputtering) in which a discharge is established in an argon plus nitrogen or carbon containing gas between the anode and cathode. The gas ions are accelerated to the cathode where they eject or sputter away the cathode material. The sputtered material is deposited on a suitably placed substrate. The reaction of the metal atoms with C or N probably occurs at the substrate. Since the cathode material has the same composition as the metal carbide or nitride to be formed, the method is limited to a fixed composition for the metal atom component of the carbide or nitride, and is ideally suited to binary carbides and nitrides.

Figure 6 shows a reactive sputtering device built by Bell et al. (35) to deposit ternary nitride films. The associated bell jar and ultrahigh vacuum system are not shown. A low pressure of 10^{-3} Torr of high purity argon plus 10^{-4} Torr N_2 is introduced into the sputtering assembly. The gas is ionized by electrons supplied by the cathode, with the anode-to-cathode voltage at about 40 V. A magnetic field is usually aligned parallel to the central axis to increase the probability of an ionizing event. When the discharge is established, the material to be deposited (target) is biased with a high negative potential. Gas ions accelerated to the target liberate (sputter) neutral target atoms. Sufficient momentum is transferred to the target atoms to propel these atoms to the substrate. By altering the potentials of the two targets shown in the figure, the atomic ratio of target materials landing on the substrate is altered.

Mitszuoka et al. (36) have discovered that asymmetric ac sputtering improves the quality of NbN films, compared to films prepared by dc glow

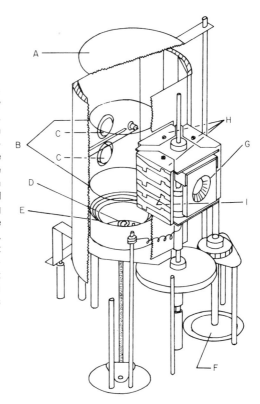

Fig. 6. Device for preparing ternary nitride thin films of Nb–Zr–N by the process of reactive sputtering. A low partial pressure of nitrogen is introduced into the bell jar assembly for the formation of the nitrides. The Nb and Zr targets are individually biased for composition control of the resultant film. Principal parts are designated by the following letters: A, virtual cathode; B, anode rings; C, targets of Nb and Zr; D, cathode; E cathode heat shields; F, cylindrical magnet to rotate substrate holder; G, deposition monitor; H, slide holders; and I, substrate and substrate mask for simultaneous deposition of four samples.

discharge techniques. The superconducting transition temperature was used to determine the quality of the films. In this technique a partially rectified ac potential is applied between two electrodes, both of which are of the same target composition. The substrate is placed on the electrode with the lower bombarding potential. Material is deposited onto the substrate during one half of the ac cycle and part of that material is sputtered away during the other half of the cycle. Because the superconducting transition temperatures of the ac films are higher than those of the dc films, Mitszuoka *et al.* believe that the ac method increases the nitrogen content of the films to make the composition more nearly stoichiometric (see Chapter 7).

The reactive sputtering method has thus far been applied to only a few binary and ternary nitrides. Nevertheless, the ease of preparation, the high quality of the resultant nitrides, and the ease of composition control indicate that this method is superior in several ways to the evaporation-reaction technique first described. Epitaxial films have also been prepared by this technique without difficulty (*38*).

V. The Problem of Characterization

Characterizing carbide and nitride samples is a difficult and challenging task not sufficiently appreciated by many investigators. In writing this text, the author was often confronted with literature reports describing a careful experiment about a property of carbides and nitrides on an uncharacterized sample. Since the properties are very sensitive to a number of factors— composition, porosity, crystal structure, etc.—the results of such experiments could not be included in this survey. Such experiments are meaningless! One cannot, however, be overly critical about characterizing carbides and nitrides because, for reasons described below, it is doubtful that many samples can be completely characterized. What should be expected is a knowledge of the characterization problem, a judicious decision about what characterization is pertinent to the experiment, and then a reasonable attempt to characterize these parameters.

The properties of carbides and nitrides are dependent upon a number of factors:

(1) crystal structure and lattice parameters.

(2) chemical composition including nonmetal-to-metal ratio, combined and free carbon concentration in carbides, and impurity concentration, especially oxygen.

(3) overall defect structure including vacancy concentration, grain size dislocation structure, porosity and distribution of porosity.

(4) sample homogeneity.

Not all of these factors may be important in measuring a particular property, although crystal structure, lattice parameters, composition, nonmetal-to-metal ratio, impurity concentration, and sample homogeneity should be characterized for nearly all property measurements. Porosity affects mechanical and electrical properties, but it may not be important in certain thermodynamic or magnetic measurements. Several techniques for characterization of each factor are discussed below.

Since most crystal structures of carbides and nitrides are known, confirmation of a suspected structure is readily performed using Debye–Scherrer techniques. This method is also useful in determining sample homogeneity by observing the sharpness of the splitting of $K_{\alpha1}$ and $K_{\alpha2}$ lines and in detecting impurity phases. The limit of detection of impurity phases, however, is usually a few percent. Under certain conditions even larger impurity concentrations escape notice. Also impurities such as oxygen form solid solutions with carbides and nitrides and therefore failure to observe an oxide impurity phase is not complete insurance that contamination has not occurred.

Careful X-ray analysis is needed to detect ordering of carbon or nitrogen atoms. Since the scattered X-ray intensity from a given atom is roughly proportional to the square of the atomic number, the positions of carbon or nitrogen in the crystal structure are not readily determined by this technique. Ordering is sometimes accompanied by a slight distortion of the host metal-atom structure, which results in a change in crystal symmetry. In these cases, X-ray analysis can be used to detect the structural change in the metal sub-structure, but the technique still cannot identify the type of ordering on the nonmetal sites. In many instances, ordering of carbon or nitrogen occurs without a symmetry change in the metal substructure and the detection of a structural change by X rays would necessitate complex Fourier analysis of the intensity spectrum.

Unambiguous determination of the type of ordering of the nonmetal atoms can be accomplished by electron or neutron diffraction. In both techniques the diffracted intensity no longer depends upon the square of the atomic number and the positions of the nonmetals can be determined from diffraction patterns. Electron diffraction is difficult, however. Thin-film sections must be prepared, a procedure complicated by the brittle and porous nature of the materials and a general lack of knowledge about thinning techniques. Neutron diffraction is ideally suited for this task, especially since intensity spectra of many ordered carbides have been previously compiled, but the technique is not readily available to most groups, and relatively large samples are needed.

Chemical analyses of carbides and nitrides are usually restricted to determinations of carbon (combined and free), nitrogen, and impurities, although in more careful work, the amount of the transition element is also directly determined. Analyses of the transition metal are not always performed because most sample preparation techniques, such as powder metallurgy, cause little loss of the transition metal. These analyses can, however, be performed to check the accuracy of the carbon or nitrogen determination provided impurities occur only in small concentrations. Amounts of carbon and nitrogen, on the other hand, can be significantly altered during processing and these quantities must be determined. In the case of carbides, particularly carbon-rich ones, not all the carbon is combined, and some exists as a second phase of free graphite or carbon; separate determinations of both the combined and free carbon are therefore necessary. Impurity determinations generally consist of a spectroscopic analysis of a selected number of suspected impurities and also analysis of oxygen content. Kriege (39) and Dutton et al. (39a) have given excellent descriptions of the techniques for reliable chemical analyses of over 25 different refractory carbides and nitrides.

The amount of combined carbon can be obtained by first analyzing for total carbon and for free carbon and then taking the difference. The total carbon content is determined by heating the carbide in a stream of oxygen; the carbide is converted into an oxide and the carbon is converted into CO_2. The amount of CO_2 is determined by absorbing it in Ascarite and monitoring the weight change of the Ascarite or by measuring the conductivity of the combusted CO_2–O_2 gas mixture as is done in the commercial Leco unit. To analyze free carbon or graphite, the carbide is dissolved in a mixture of hydrofluoric and nitric acid. Free carbon or graphite is not dissolved by the acid and forms a residue which is collected, washed, dried, and then converted by combustion to CO_2 for final determination. With good calibration techniques, the accuracy in analyzing for total carbon content is about 0.05 wt %; the accuracy for free carbon is considerably less. The lower accuracy in the free carbon analyses is due in part to a smaller percentage of the sample that is in the form of free carbon, due to the formation of tars and the loss of finely divided graphite through the filter and also possibly due to the loss of free carbon because it is in an activated state (40).

Chemical analysis for nitrogen in nitrides is a difficult problem. Most commercial laboratories use the Kjeldahl technique which is well suited for nitrogen determinations in organic materials but which gives low and unreliable results for transition-metal nitrides. Better results are obtained with a Dumas method (39, 41, 42), modified to ensure complete decomposition of the nitride. In the conventional Dumas method (41) fine nitride powders are mixed with CuO powder. The powders are placed in a tube furnace which is flushed with CO_2 and then heated to about 1000°C. The CuO reacts with the nitride to form a transition-metal oxide and in the process N_2 is liberated. The CO_2–N_2 gas mixture is then flushed out of the furnace with additional CO_2 and passed through a KOH solution where the CO_2 is absorbed and the N_2 measured volumetrically. Because of the refractory nature of transition-metal nitrides the method is modified to aid the decomposition by adding oxides of low melting point such as PbO, to the CuO–nitride mixture.

Because of the general unavailability of the Dumas technique on a commercial scale and because of the necessity of modifying this technique for refractory nitrides, many experimenters have developed their own special techniques. Unfortunately, no standard samples exist to check the accuracy of these techniques, and so, variations between laboratories of a few atomic percent of nitrogen are not uncommon. Thus the best accuracies available to date on nitrogen content in nitrides is about 1–2 at. %.

Oxygen is the most difficult impurity to eliminate and one of the most difficult to analyze. Its effect on properties, however, is usually significant; therefore its presence cannot be ignored. Oxygen forms a solid solution with both carbides and nitrides in the form Me(C, O) or Me(N, O). Once oxygen is dissolved in the carbide or nitride, it is difficult—and in some cases nearly impossible—to remove. Fifth and sixth-group nearly stoichiometric carbides can be purified by heating under high vacuum conditions at elevated temperatures and in the presence of excess carbon. The excess carbon reacts with the oxygen to form CO, which is then removed by the vacuum system. Nearly stoichiometric fourth-group carbides can be purified by this method only if the temperature is close to the melting point and if the vacuum is better than 10^{-6} Torr. Nonstoichiometric carbides and subcarbides are particularly difficult to purify since the CO pressure, which depends on carbon and oxygen activities, drops rapidly in the unsaturated phases.

The best technique for sample preparation is to avoid all possible contamination. Direct reaction of metal or metal hydrides with C or N is preferred to using metal oxide powders as starting materials. Fourth-group carbides and nitrides should not be heat treated in vacuums of less than 10^{-6} Torr. These phases are best treated in high-purity inert gases.

If carbides and nitrides become contaminated with oxygen, it is difficult to apply normal analytical techniques. Oxygen in metals is usually analyzed by the vacuum-fusion technique. The sample is fused under vacuum in an inductively heated graphite crucible, sometimes with the aid of a flux of molten metal. The gaseous products are collected with a diffusion pump and analyzed; the oxygen from the sample appears primarily as CO. To fuse refractory metals and refractory carbides and nitrides, very high temperatures are necessary, and special techniques are used, such as fusion in a Pt bath (43) and fusion in a graphite mold at 2400–2800°C (44). For the fourth-group metals and carbides these techniques are only partially successful. Stability of oxygen in the fourth group is great, and therefore oxygen is difficult to remove. Neutron activation analysis has been successfully used to analyze for oxygen in certain carbides. Oxygen is activated through the reaction $^{16}O(n,p)^{16}N$. The amount is determined by monitoring the 6.1 and 7.1 MeV γ-radiation from the ^{16}N (45).

Characterizing the defect structure in carbides and nitrides is another complex problem. Some types of defects such as porosity, distribution of porosity, and grain size can be characterized by a combination of metallography and density measurements. Special metallographic techniques are required for each carbide or nitride system. The hard materials are polished either with diamond powder or micron-size alumina powder in an aqueous solution of Murakami's reagent $\{K_3Fe(CN)_6 + KOH\}$. Because of their high degree of chemical stability very strong etching solutions (HF, HNO_3, and H_2O_2) are required. Furthermore, the proper etching solution may change with the nonmetal-to-metal ratio. Successful techniques and solutions for different systems have not been compiled, but the extensive research by Rudy and co-workers[2] on carbides is a valuable starting place to develop these metallographic techniques.

Determining the vacancy concentration is difficult because of inaccuracies in determining the exact overall chemical composition. Even at stoichiometry appreciable vacancy concentrations can exist on both metal and nonmetal lattice sites. In TiN the vacancy concentration at stoichiometry on both sublattices may be as high as 4 at. % (46). The vacancy concentration can be determined by comparing X-ray density with experimentally determined density. If the chemical composition is not exactly known, however, very large errors in estimating the vacancy concentrations will be encountered, because the error in composition is added or subtracted to the observed vacancy concentration. This problem is particularly difficult in the case of the mononitrides, where nitrogen content can be analyzed only to within 1%.

Sample homogeneity, particularly in the nonstoichiometric carbides and nitrides, can be a problem because of the slow diffusion rates of carbon and nitrogen. To eliminate inhomogeneities, nonstoichiometric nitrides should be heated in a partial pressure of nitrogen of the final desired composition. The sample can then adjust its nitrogen composition to an equilibrium value. Sometimes, however, the partial pressure is too low to allow homogenization and equilibration to occur in a reasonable amount of time or with a reasonable flow rate of gas. Also, the partial pressure for a particular composition may be inexactly known. In these cases, an inhomogeneous starting material can be heated in a static atmosphere at elevated temperatures. The nitrogen partial pressure of the atmosphere and the sample composition will adjust themselves to equilibrium values and homogenization can proceed to completion. The sharpness of Debye–Scherrer X-ray patterns can be used to indicate the homogeneity; some care should be taken to sample properly several different parts of the specimen.

References

1. R. Kieffer and P. Schwarzkopf, in collaboration with F. Benesovsky and W. Leszynski, "Hartstoffe und Hartmetalle." Springer, Vienna, 1953.
2. P. Schwarzkopf and R. Kieffer, in collaboration with W. Leszynski and F. Benesovsky, "Refractory Hard Metals." Macmillan, New York, 1953.

[2] E. Rudy et al. have published metallographic techniques in several volumes of AFML-TR-65-2, Air Force Materials Laboratory, Research and Technology Division, Wright-Patterson Air Force Base, Ohio.

3. J. F. Lynch, C. G. Ruderer, and W. H. Duckworth, "Engineering Properties of Selected Ceramic Materials." Am. Ceram. Soc., Columbus, Ohio, 1966.
4. E. K. Storms, "The Refractory Carbides." Academic Press, New York, 1967.
5. H. J. Goldschmidt, "Interstitial Alloys." Plenum Press, New York, 1967.
6. H. L. Schick, ed., "Thermodynamics of Certain Refractory Compounds," Vols. 1 and 2. Academic Press, New York, 1966.
7. P. T. B. Shaffer, "High-Temperature Materials, Materials Index." Plenum Press, New York, 1964.
8. G. V. Samsonov, "High-Temperature Materials, Properties Index." Plenum Press, New York, 1964.
9. G. V. Samsonov and Ya. S. Umanskiy, "Tverdyye Soyedineniya Tugoplavkikh Metallov." State Sci.-Tech. Lit. Publ. House, Moscow, 1957; for English translation, see *NASA Tech. Trans.* F-102 (1962).
10. H. H. Hausner and M. G. Bowman, ed., "Fundamentals of Refractory Compounds." Plenum Press, New York, 1968.
11. R. Kieffer and F. Benesovsky, "Encyclopedia of Chemical Technology," 2nd ed., Vol. 4, p. 70, and Vol. 13, p. 814. Wiley (Interscience), New York, 1964.
12. R. G. Lye, G. E. Hollox, and J. D. Venables, in "Anisotropy in Single-Crystal Refractory Compounds" (F. W. Vahldiek and S. A. Mersol, eds.), Vol. 2, p. 445. Plenum Press, New York, 1968.
13. G. E. Hollox, *Mater. Sci. Eng.* 3, 121 (1968–1969).
14. P. M. McKenna, *Metal Progr.* 36, 152 (1939).
15. H. Pfau and W. Rix, *Z. Metallk.* 45, 116 (1954).
16. J. Corteville and L. Pons, *C. R. Acad. Sci.* 260, 4477 (1965).
17. T. Takahashi and E. J. Freise, *Phil. Mag.* [8] 12, 1 (1965).
18. A. P. Gerk and J. J. Gilman, *J. Appl. Phys.* 39, 4497 (1968).
19. A. D. Kiffer, WADD T.R.6 0–52 (1960); Defense Doc. Center No. AD-238-061.
20. W. Precht and G. E. Hollox, *J.Cryst.Growth* 3/4, 818 (1968); see also RIAS Rep. 68-9c (1968).
21. J. M. Tobin and L. R. Fleischer, Westinghouse Electric Co., Pittsburgh Pa., Astronuclear Lab. Tech. Rep. AF 33 (615) 3982; AFML-TR-67-137, Part I (1967); Defense Doc. Center No. AD-663-248 (1968); also AFML-TR-67-137, Part II.
22. A. Brager, *Acta Physicochim.* URSS 10, 593 (1939); 11, 617 (1939).
23. K. Moers, *Z. Anorg. Allg. Chem.* 198, 243 (1931).
24. N. V. Troitskaya and Z. G. Pinsker, *Sov. Phys.—Crystallogr.* 4, 33 (1960); 6, 34 (1961); 8, 441 (1964).
25. V. I. Khitrova and Z. G. Pinsker, *Sov. Phys.—Crystallogr.* 6, 712 (1962).
26. N. Schwartz, "Vacuum Symposium Transactions," p. 325. Macmillan, New York, 1963.
27. D. Gerstenberg and C. J. Calbick, *J. Appl. Phys.* 35, 402 (1964).
28. D. Gerstenberg, *J. Electrochem. Soc.* 113, 542 (1966).
29. E. Krikorian and R. J. Sneed, *J. Appl. Phys.* 37, 3674 (1966).
30. D. Gerstenberg and P. M. Hall, *J. Electrochem. Soc.* 11, 936 (1964).
31. L. I. Maissel and P. M. Schaible, *J. Appl. Phys.* 36, 237 (1965).
32. F. Vratny and N. Schwartz, *J. Vac. Sci. Technol.* 1, 79 (1964).
33. D. A. McLean, N. Schwartz, and E. D. Tidd, *Proc. IEEE* 52, 1450 (1964).
34. J. Sosniak, *J. Vac. Sci. Technol.* 4, 87 (1967).
35. H. Bell, Y. M. Shy, D. E. Anderson, and L. E. Toth, *J. Appl. Phys.* 39, 2797 (1968).
36. T. Mitszuoka, T. Yamashita, T. Nakazawa, Y. Onodera, Y. Saito, and T. Anayama, *J. Appl. Phys.* 39, 4788 (1968).

37. The following theses should be consulted: H. Bell, M. S. Thesis, "An Investigation of Reactively Sputtered Thin Film Superconducting Nitrides," University of Minnesota, 1966; Y. M. Shy, M. S. Thesis, "Superconducting Critical Magnetic Fields, Current Densities and Temperatures in Bulk and Thin Film Samples of the Ternary Nb_xTi_yN System," University of Minnesota, 1967.

38. Y. M. Shy and L. E. Toth, private communication (1969).

39. O. H. Kriege, LA-2306 (1959).

39a. R. E. Dutton, G. J. McKinley, D. McLean, and H. F. Wendt Union Carbide Tech. Rep. C-29 (1965); Defense Doc. Center No. AD-459-048.

40. E. K. Storms, private communication (1969).

41. W. F. Hillebrand and G. E. F. Lundell, "Applied Inorganic Analysis." Wiley, New York, 1929.

42. W. Kern and G. Brauer, *Talanta* **11**, 1177 (1964).

43. W. G. Smiley, *Anal. Chem.* **27**, 1098 (1955).

44. M. E. Smith, J. M. Hansel, R. B. Johnson, and G. R. Waterbury, *Anal. Chem.* **35**, 1502 (1963).

45. D. Taylor, "Neutron Irradiation and Activation Analysis." Van Nostrand, Princeton, New Jersey, 1964.

46. P. Ehrlich, *Z. Anorg. Allg. Chem.* **259**, 1 (1949).

2

Crystal Chemistry

I. Introduction

Compounds of carbon and nitrogen with transition metals have structures that may be simply described as close-packed—or nearly close-packed—arrangements of metal atoms with smaller nonmetal atoms inserted into interstitial sites. For most structures, there are no apparent carbon–carbon or nitrogen–nitrogen localized interactions that "mold" structures as one might expect in analogy to organic compounds. Borides, however, do clearly exhibit localized interactions between boron atoms in their structures, including chains, layers, or three-dimensional networks that influence the geometry of the structure (*1*). An important characteristic of carbides and nitrides is the metal–nonmetal interaction and the geometry of the interstitial site. Carbon and nitrogen are normally located in either an octahedral interstitial site or in the center of a trigonal prism. Figure 1 shows the types of interstitial sites in fcc, bcc, hcp, and simple hexagonal structures. The interstitial atom and its nearest metal neighbors comprise a structural unit (coordination polyhedron) with which the entire structure may be conveniently reconstructed. One may view the structures as either a metal structure with occupied interstitial sites or as a structure composed primarily of coordination polyhedra.

There are a number of advantages to either type of description. The interstitial model is often the easiest to visualize and the most familiar. Metal

29

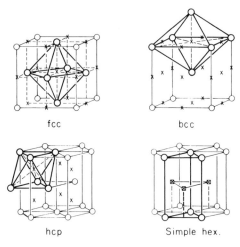

fcc

bcc

Fig. 1. Octahedral interstitial sites in fcc, bcc, and hcp structures and trigonal prism interstitial sites in the simple hexagonal structure. ○ : metal atoms ; × : octahedral interstitial site ; ⊠ : trigonal prism interstitial site.

hcp

Simple hex.

atoms in carbides and nitrides often form simple fcc, hcp, or simple hexagonal substructures. Filling all the octahedral interstitial sites in fcc results in the $B1$(NaCl) structure. The metal substructure may also be described according to the sequential ordering of the close-packed metal-atom layers with the usual designations: —ABCABC— for fcc, —AB,AB— for hcp, and —AA— for simple hexagonal. Carbide and nitride structures have the simple notations —AXBX'CX'', AXBX'CX''— for $B1$, —AXBX,AXBX— for L_3', and —AX,AX— for the WC type, where X represents a complete or partial filling of the interstitial sites (octahedral or trigonal prism) created by two adjacent metal layers. X' and X'' differ from X layers by a lateral shift.

The principal advantage of the coordination polyhedra model is for describing complex crystal structures in which transition atoms no longer occupy close-packed or nearly close-packed positions but in which well defined coordination polyhedra exist. In many ways the crystal chemistry of carbides and nitrides can be developed from the viewpoint of coordination polyhedra in the same way that silicates are described in terms of SiO_4 tetrahedra, and aluminum hydroxide and related compounds are described in terms of $Al(OH)_6$ octahedra.

Throughout this chapter, designations for crystal structure types conform to those used by Pearson (*2, 2a*) in *A Handbook of Lattice Spacings and Structures of Metal and Alloys*. Pearson has devoted a special section of his book to carbides and nitrides; this section includes information about lattice parameters, space groups, atomic positions, and Strukturbericht symbols. According to Pearson, "the best one can do in classifying structures is to name each structure type after a representative substance (*2, 2a*)."

This procedure was generally followed, although the familiar Strukturbericht symbols (or some further identifying information) were used for the more common and simple structures. In some cases the literature on carbides and nitrides frequently refers to a structural name other than that used by Pearson, such as TiP or γ'-MoC for AsTi(B_i). Here, both names were used, although the more common one was preferred. Many of the phases whose crystal structures are described here are prefixed by a Greek letter usually to denote the existence of a high or low-temperature crystal form. These Greek-letter designations also conform to those used by Pearson and to the phase diagram designations described in the next section.[1]

II. Hägg Compounds

IIA. Hägg Rules

In 1931, Hägg (3) formulated an interesting set of empirical rules regulating the crystal structure types formed by transition-metal carbides, nitrides, borides, and hydrides. Although minor exceptions to the rules have been found, their general features remain valid. According to Hägg, the structure of transition-metal carbides, nitrides, borides, and hydrides is determined by the radius ratio $r = r_X/r_{Me}$. Here r_X and r_{Me} are the radius of the interstitial and transition-metal atom, respectively. If r is less than 0.59, the metal atoms form very simple structures: $A1$, $A2$, $A3$, or simple hexagonal. If r is greater than 0.59, the transition metal and interstitial elements form complicated structures. When r is less than 0.59, the light elements are accommodated in the largest interstitial sites of the relatively simple metal host structure. The interstitial site must necessarily be smaller than the interstitial atom that it is to accommodate; otherwise there will be insufficient bonding between the metal and nonmetal atoms and the structure will become unstable. On the other hand, the interstitial site cannot be too much smaller than the interstitial atom; otherwise the presence of the interstitial atom will expand the metal host lattice to the point where the metal–metal interactions will become weak and the structure will lose its stability. Hägg observes that the atomic volume occupied by each metal atom is less in the complex structures than it would be in a hypothetical simple structure if r is greater than 0.59. Complex structures form in order not to allow undue

---

[1] Parthé and Yvon have recently proposed an alternate notation for carbide structures. See E. Parthé and K. Yvon, *Acta Cryst.* **26B**, 153 (1970) and K. Yvon and E. Parthé, *Acta Cryst.* **26B**, 149 (1970).

TABLE I

METAL LATTICES AND INTERSTITIAL SITES FOR HÄGG'S "SIMPLE" CRYSTAL STRUCTURES

Metal structure	Metal atomic positions	Interstitial site	Interstitial atomic positions	Metal–nonmetal neighbors	Minimum r for occupancy	Maximum r for occupancy
fcc	000, $0\frac{1}{2}\frac{1}{2}$, $\frac{1}{2}0\frac{1}{2}$, $\frac{1}{2}\frac{1}{2}0$	Octahedral	$\frac{1}{2}\frac{1}{2}\frac{1}{2}$, 100 $0\frac{1}{2}0$, $00\frac{1}{2}$	6	0.41	0.59
		Tetrahedral	$\pm\frac{1}{4}\frac{1}{4}\frac{1}{4}$, $\pm\frac{1}{4}\frac{3}{4}\frac{3}{4}$ $\frac{3}{4}\frac{1}{4}\frac{3}{4}$, $\frac{3}{4}\frac{3}{4}\frac{1}{4}$	4	0.23	—
hcp	000, $\frac{1}{3}\frac{2}{3}\frac{1}{2}$	Octahedral	$\frac{2}{3}\frac{1}{3}\frac{1}{4}$, $\frac{2}{3}\frac{1}{3}\frac{3}{4}$	6	0.41	0.59
		Tetrahedral	$00\frac{3}{8}$, $\frac{1}{3}\frac{2}{3}\frac{7}{8}$ $00\frac{5}{8}$, $\frac{1}{3}\frac{2}{3}\frac{1}{8}$	4	0.23	—
bcc	000, $\frac{1}{2}\frac{1}{2}\frac{1}{2}$	Tetrahedral	$\pm0\frac{1}{2}\frac{1}{4}$, $\pm\frac{1}{2}0\frac{1}{4}$ $\pm\frac{1}{2}\frac{1}{4}0$, $\pm0\frac{1}{4}\frac{1}{2}$ $\pm\frac{1}{4}\frac{1}{2}0$, $\pm\frac{1}{4}0\frac{1}{2}$	4	0.29	—
Simple hexagonal $c/a = 1$	000	Trigonal prism	$\pm\frac{1}{3}\frac{2}{3}\frac{1}{2}$	6	0.53	0.59

TABLE II

RADIUS RATIOS FOR TRANSITION-METAL ATOMS[a]

Element	Sc	Ti	V	Cr	Mn	Fe	Co	Ni
Atomic radius CN (12) Å	1.620	1.467	1.338	1.267	1.261	1.260	1.252	1.244
C/Me radius ratio	0.467	0.526	0.576	0.609	0.611	0.612	0.616	0.620
N/Me radius ratio	0.457	0.504	0.553	0.584	0.587	0.587	0.591	0.595

Element	Y	Zr	Nb	Mo	Tc	Ru	Rh	Pd
Atomic radius CN (12) Å	1.797	1.597	1.456	1.386	—	1.336	1.342	1.373
C/Me radius ratio	0.429	0.483	0.530	0.556	—	0.577	0.574	0.561
N/Me radius ratio	0.418	0.463	0.508	0.534	—	0.554	0.551	0.539

Element	La	Hf	Ta	W	Re	Os	Ir	Pt
Atomic radius CN (12) Å	1.871	1.585	1.457	1.394	1.373	1.350	1.355	1.385
C/Me radius ratio	0.412	0.486	0.529	0.553	0.561	0.571	0.569	0.557
N/Me radius ratio	0.396	0.467	0.508	0.531	0.539	0.548	0.546	0.534

[a] Pauling's atomic sizes are used (2, 2a, 5).

dilation of the simple metal structure by relatively large interstitial atoms. Parthé (4) has explained the occurrence of the simple metal structures of the Hägg compounds in a more quantitative manner using his space filling theory.

The types of interstitial sites in simple metal structures according to Hägg are listed in Table I (see also Fig. 1). Values of r for the transition elements and the type of interstitial site occupied by C and N are listed in Table II (2a, 5). For the transition elements, the tetrahedral interstitial sites of the simple metal structures are too small to accommodate C or N, and only the octahedral and trigonal prism interstitial sites are occupied. Occupation of all the octahedral interstitial sites in a fcc metal lattice results in the $B1$(NaCl) structure, which is very common among monocarbides and mononitrides. A random occupation of half of the interstitial sites in the hcp metal sub-structure results in the L_3' structure common among the Me_2C and Me_2N structures. Occupation of the trigonal prism interstitial sites in the simple hexagonal structure results in the WC structure.

The crystal structures of the carbide and nitride interstitial compounds are entirely different from those of the terminal interstitial solid solutions of C or N with a transition-metal element. Andrews and Hughes (6) noticed an unusual correlation between the crystal structure of the metal element and the crystal structures of its carbides and nitrides. With one exception, cobalt, the crystal structure of the metal changes upon forming the carbide and nitride compounds. If the element has an hcp structure, it will not form a carbide or nitride in which the metal substructure is hexagonal. If the element has a fcc structure, it will not form a carbide and nitride in which the metal substructure is cubic. If the element has the bcc structure and no close-packed allotropes, it will form carbides and nitrides in which the metal atoms form both hexagonal and cubic substructures. These relation-ships are shown in Table III. It is clear from Table III that crystal structures of carbides and nitrides are not solely a function of size relationships, but are also dependent upon the stabilizing effect of the interstitial in a particular coordination polyhedron and also upon metal–metal interactions. These considerations will be discussed in a later chapter on bonding.

In the bcc structure there are two types of interstitial sites. The larger one is the tetrahedral site situated at $\frac{1}{2}\,\frac{1}{4}\,0$ and equivalent sites, and the smaller one is a distorted octahedral site at $0\,0\,\frac{1}{2}$ and $\frac{1}{2}\,\frac{1}{2}\,0$ and equivalent sites. Since the interstitial atoms C and N are difficult to accommodate in these sites and since there appears to be a preference for an octahedral site by the nonmetal, the metal atoms form a fcc or hcp substructure to provide the larger octahedral site. The small size of the distorted octahedral site in bcc may also account for the limited solubility of C and N in many IV- to VI-group elements.

TABLE III

METAL AND INTERSTITIAL ALLOY STRUCTURES[a]

Group	Element	Metal structures		Cubic substructure for metal atoms in carbide and nitride	Hexagonal sub-structure for metal atoms in carbide and nitride
		bcc	hcp		
	Ti	E[b]	E	E	—
IV	Zr	E	E	E	—
	Hf	E	E	E	—
	V	E	—	E	E
V	Nb	E	—	E	E
	Ta	E	—	E[c]	E
	Cr	E	—	E	E
VI	Mo	E	—	E	E
	W	E	—	E	E

[a] See Andrews and Hughes (6).

[b] The letter "E" denotes the existence of the phase with the designated structure, "—" denotes phase does not exist with that structure. No phases with fcc structure exist.

[c] TaN with a cubic B1 structure exists in thin films.

IIB. Crystal Structures of Interstitial Carbides and Nitrides

According to Table II, most transition-metal carbides and nitrides have r values of less than 0.59. Therefore, by the Hägg rules, the metal atoms should form simple structures in which they are always in close-packed layers. Table IV lists the various sequences of layer stacking for the metal atoms, and the complete interstitial sequential ordering in carbides and nitrides (2, 2a, 7–9). Here the X, X', and X" refer to carbon or nitrogen layers and the primes are used in the same sense as the letters A, B, and C of the metal-atom layers to denote a lateral shift of one layer with respect to the other layers (10). Representative compounds crystallizing in these structures are also given in Table IV.

Figure 2 illustrates the sequential ordering of the phases listed in Table IV. The octahedral interstitial site occurs in alternating sequences of close-packed metal atom layers, i.e., AB or BC, etc. The trigonal prism interstitial site occurs when the close-packed metal atom layers are not laterally displaced, i.e., AA or BB. The WC type is an example of a structure containing the trigonal prism interstitial site; the NaCl(B1), L_3', anti-NiAs and

TABLE IV

INTERSTITIAL CARBIDES AND NITRIDES

Stacking sequence of metal layers	Complete stacking sequence	Crystallographic designation	Ideal composition	Representative phases among carbides and nitrides[a]
A, A …	AX, AX …	WC type (B_h)[b]	MeX	WC, γ-MoC, $Ti_{0.7}Ni_{0.3}N$, $Ti_{0.7}Co_{0.3}N$, $Mo_{0.8}Ni_{0.2}N_{0.9}$, $Mo_{0.8}Co_{0.2}N_{0.9}$, δ-$TaN_{0.8-0.9}$, δ-WN
AABB, AABB …	AXAX'BX'BX', AXAX'BX''BX' …	B_i TiP or γ'-MoC type	MeX	γ'-MoC, ϵ-NbN[a]
ABC, ABC …	AXBX'CX'', AXBX'CX'' …	B1 (NaCl)	MeX	CrN, LaN, YN, ScN, TiN, VN, ZrN, HfC, ZrC, TiC, TaC, NbC, δ-NbN, α-WC, α-MoC, HfN, ScC (VC ?)
AB, AB …	AXBX, AXBX …	$B8_1$ anti-NiAs type	MeX	δ'-NbN
AB, AB …	AXBX, AXBX, …	L_3' (W_2C)	Me_2X	β-Mo_2C, β-W_2C, β-Ta_2C, β-V_2C, β-Nb_2N, γ-Ta_2N, γ-Nb_2C, Fe_2N, ϵ-Mn_2N
AB, AB …	AXBX', AXBX' …	L_3' derivatives	Interstitials ordered	See Table VI
ABCACB, ABCACB …	AXBX'CX''AX'''CX'BX, AXBX'CX''AX'''CX'BX, …	η-MoC ($hP10$)	MeX	η-MoC (Mo_3C_2)

[a] Notation for Nb–N phases follows that of Brauer and Esselborn (9).

[b] WC has the following atomic positions according to neutron diffractions results (7, 8): 1 W at 0 0 0, 1 C at $\frac{1}{3}\frac{2}{3}\frac{1}{2}$. It was thought for some time that WC had the B_h structure in which one carbon would be random in the $\frac{1}{3}\frac{2}{3}\frac{1}{2}$ and $\frac{2}{3}\frac{1}{3}\frac{1}{2}$ sites. There have not been neutron diffraction studies on the other carbides and nitrides belonging to this group to determine if they are WC or B_h type.

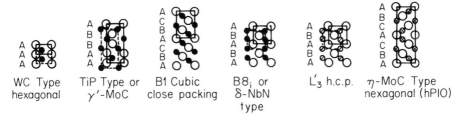

WC Type hexagonal	TiP Type or γ'-MoC	B1 Cubic close packing	B8ᵢ or δ-NbN type	L'₃ h.c.p.	η-MoC Type hexagonal (hP10)

Fig. 2. Sequential ordering of metal-atom layers in the crystal structures of interstitial carbide and nitride compounds. ●: Carbon sites; ⊘: Carbon statistically on $\frac{1}{2}$ of the sites; ⊗: Carbon statistically on $\frac{2}{3}$ of the sites; ○ Me atoms.

η-MoC(hP10) types are examples of structures with octahedral interstitial sites. The TiP type (AsTi-B_i) contains both octahedral and trigonal prism interstitial sites.

Jellinek (*11*) has clarified the structural relationships between the WC, NaCl, anti-NiAs, and L_3' types by relating these structures to the NiAs

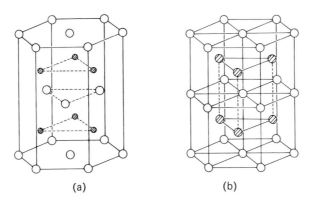

(a)   (b)

Fig. 3. (a) Nickel Arsenide structure. ○: As; ⊘: Ni. (b) WC-type structures (six-unit cells). ○: W; ⊘: C.

structure which forms an important link between nonmetallic compounds (ionic or homopolar) and metallic compounds. The NiAs structure is shown in Fig. 3a. In the NiAs structure, Ni forms a simple hexagonal substructure, and As forms an hcp substructure. Each Ni atom is surrounded by six As atoms in an octahedral coordination, and each As atom is surrounded by six Ni atoms in a trigonal prism coordination. In WC, both W and C form simple hexagonal substructures (Fig. 3b). In the anti-NiAs structures, the metal atoms form in hcp and the nonmetal atoms form in simple hexagonal;

that is the metal and nonmetal atoms have exchanged relative positions. Only δ'-NbN has the anti-NiAs structure, with nearly all the octahedral sites filled with nitrogen atoms; a small percentage of the sites are vacant because of a stoichiometry deviation. Many carbides and nitrides, however, because of the small r_X/r_{Me} ratio have anti-NiAs structural derivatives such as L_3' and η-MoC type. In the L_3' structures, the metal atoms still are hcp, but only half of the octahedral interstitial sites are occupied in a random manner. In many Me_2X structures the nonmetal atoms order. The NaCl structure does not belong to the anti-NiAs class because the former is a fcc-type structure where both the metal and nonmetal atoms are in octahedral coordination about each other. The WC type is considerably different from the NaCl type in that the coordination polyhedra have changed from octahedra to trigonal prisms. These relationships are summarized in Table V. Figure 4 shows a modified version of the Jellinek diagram which pictorially

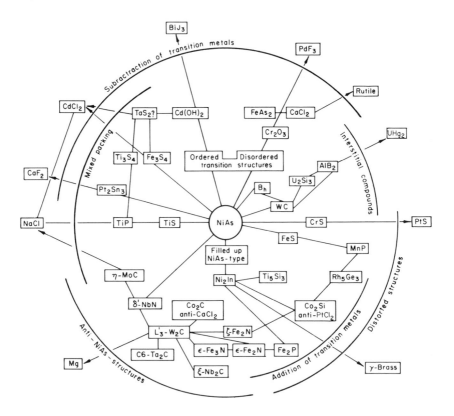

Fig. 4. A modified Jellinek diagram showing the central relationship of the NiAs structure between the WC, NaCl, δ'-NbN, and other type structures.

TABLE V

SUBSTRUCTURES AND COORDINATION POLYHEDRA OF SEVERAL HÄGG PHASES RELATED TO THE NiAs-TYPE STRUCTURE

Type	Atomic positions of metal atoms	Atomic positions of nonmetal atoms (metalloid)	Structure formed by metal atoms	Structure formed by nonmetals	Coordination of metal atoms about each nonmetal atom	Coordination of nonmetal atoms about each metal atom
NiAs	$000, 00\frac{1}{2}$	$\frac{1}{3}\frac{2}{3}\frac{1}{4}, \frac{2}{3}\frac{1}{3}\frac{3}{4}$	Simple hex	hcp	Trigonal prism	Octahedral
WC	000	$\frac{1}{3}\frac{2}{3}\frac{1}{2}$	Simple hex	Simple hex	Trigonal prism	Trigonal prism
δ'-NbN (anti-NiAs)	$\frac{1}{3}\frac{2}{3}\frac{1}{4}, \frac{2}{3}\frac{1}{3}\frac{3}{4}$	$000, 00\frac{1}{2}$	hcp	Simple hex	Octahedral	Trigonal prism
L_3'	$\frac{1}{3}\frac{2}{3}\frac{1}{4}, \frac{2}{3}\frac{1}{3}\frac{3}{4}$	1 random at $000, 00\frac{1}{2}$	hcp	Simple hex (partial)	Octahedral	Trigonal prism (partial)
NaCl	$000, \frac{1}{2}0\frac{1}{2}$ $\frac{1}{2}\frac{1}{2}0, 0\frac{1}{2}\frac{1}{2}$	$\frac{1}{2}\frac{1}{2}\frac{1}{2}, \frac{1}{2}00$ $0\frac{1}{2}0, 00\frac{1}{2}$	fcc	fcc	Octahedral	Octahedral

describes these relationships. Several of the ordered Me_2X structures, derived from the L_3' type are not included.

Fourth-group carbides and nitrides crystallize only in the NaCl structure. The composition of these phases extends over a considerable range. For example, TiC_{1-x} is stable from $TiC_{0.97}$ to $TiC_{0.50}$. Fifth-group carbides crystallize in the L_3' structure in addition to the NaCl structure. Fifth-group nitrides and sixth-group carbides crystallize in a considerable diversity of structures containing both the octahedral and trigonal prism interstitial sites. It is important to understand this competition between structures providing each type of interstitial site and to understand the reasons for the occurrence of both types in fifth-group nitrides and sixth-group carbides.

The presence of the nonmetal in the close-packed metal structure expands the distance between metal–metal contacts by a few percent over the distance in the bcc structure of the elements. The WC structure allows the metal atoms to be in closer contact with one another than they are in the NaCl or other close-packed metal structures. There are, however, fewer Me–Me contacts in the WC structure. In the NaCl structure the distance between the metal atoms is $\sqrt{2}$ times the distance between the metal and nonmetal atoms. In the WC structure with $c/a = 1$ (the approximate value for most interstitial carbides and nitrides), $D_{Me-Me} = 1.32 D_{Me-X}$. It is interesting that the WC structure is stable for the sixth-group carbides and fifth-group nitrides. The individual bond strengths of the metal atoms as estimated from the melting points of the elements are greatest for these elements. An exact comparison of the competition of these structures would involve a calculation of their respective band structures. These calculations have not been performed for the WC structure, but calculations on the NaCl type do indicate that the antibonding portion of the bands are being populated in the sixth-group carbides and that other structures might therefore be more favorable. A discussion of the band structure calculation for the NaCl phases is given in Chapter 8.

IIC. Ordering of Carbon or Nitrogen in the Hägg Compounds

In Hägg compounds in which the metal structure is either hcp or fcc, there is one octahedral interstitial site per metal atom, so that complete filling of all octahedral sites with C or N would result in a chemical formula MeX. When a substantial fraction of the interstitial sites are not occupied, ordering of the interstitial atoms can and does occur. Ordering has been found in both the substoichiometric MeX and in the Me_2X phases.

Neutron diffraction is the principal tool for determining ordering. X-ray analysis is difficult because of the large difference in scattering powers between C and N and most transition-metal atoms. Evidence for ordering of the interstitials has also been gained from phase diagram studies, differential thermal analysis (DTA), and, most recently, nuclear magnetic resonance

TABLE VI

ORDERING Me_2X PHASES

Phase	Probable high temperature modification	Low temperature modification	Stabilizing conditions	References	Comments[a]
V_2C	L_3' (random)	ζ-Fe_2N		(a, b, c, d)	Neutron
		ϵ-Fe_2N (tentative)	Possibly stabilized by oxygen	(d)	
Nb_2C	L_3' (random)	ϵ-Fe_2N	Possibly stabilized by oxygen	(d, e)	Neutron
		ζ-Nb_2C		(d)	Neutron
Ta_2C	L_3' (random)	C6		(c)	Neutron
Mo_2C	L_3' (random)	ζ-Fe_2N		(f, g)	Neutron
W_2C	L_3' (random)	C6		(h, i)	
V_2N_{1-x}	?	ϵ-Fe_2N		(d)	
$V_2C_{0.5}N_{0.5}$	?	ζ-Fe_2N		(d)	

[a] "Neutron" means that the crystal structures were determined with the aid of neutron diffraction. The structures so identified are much more certain than those determined with the aid of X rays or by other means.

References for Table VI

(a) K. Yvon, W. Rieger, and H. Nowotny, *Monatsh. Chem.* **97,** 689 (1966).

(b) N. M. Volkova and P. V. Gel'd, *Izv. Vyssh. Ucheb. Zaved. Tsvet. Met.* **3,** 77 (1965).

(c) A. L. Bowman, T. C. Wallace, J. L. Yarnell, R. G. Wenzel, and E. K. Storms, *Acta Cryst.* **19,** 6 (1965).

(d) K. Yvon, H. Nowotny, and R. Kieffer, *Monatsh. Chem.* **98,** 34 (1967).

(e) N. Terao, *J. Appl. Phys.* (*Tokyo*) **3,** 104 (1964).

(f) E. Parthé, and V. Sadagopan, *Acta Cryst.* **16,** 202 (1963).

(g) E. Rudy, S. Windisch, A. J. Stosick, and J. R. Hoffman, *Trans. AIME* **239,** 1247 (1967).

(h) L. N. Butorina, and Z. G. Pinsker, *Sov. Phys.—Crystallogr.* **5,** 560 (1960).

(i) E. Rudy, S. Windisch, and J. R. Hoffman, AFML–TR–65–2, Part I, Vol. VI (1966).

(NMR) studies. Many of the designated ordered structures for carbides and nitrides discussed in this section are still tentative. The uncertainty exists primarily in those cases in which the structural type was deduced without the aid of neutron diffraction. The high temperature crystal structure for the Me_2C phases, for example, has been tentatively assigned the L_3' structure in which carbon atoms are disordered. The assignment is based upon the consideration that the entropy factor should increase at high temperatures. The presence of this structural type, however, has not been generally confirmed with neutron diffraction. Several of the ordered types may also be influenced by impurities such as oxygen or hydrogen. Differences between one structure and another stem primarily from differences in second nearest neighbor arrangements (carbon–carbon arrangements), and hence the energy change

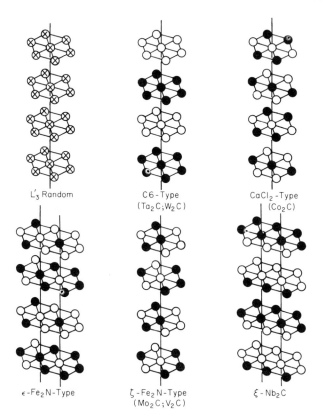

Fig. 5. Ordering of carbon or nitrogen in the Me_2X structures. The carbon or nitrogen layers are drawn perpendicular to the c-axis in the parent L_3' structure. ⊗ : interstitial atoms random at $\frac{1}{2}$ of sites; ● interstitial atoms; ○ : interstitial site vacant.

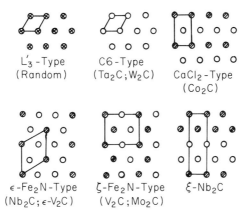

L'_3-Type
(Random)

C6-Type
(Ta$_2$C;W$_2$C)

CaCl$_2$-Type
(Co$_2$C)

ϵ-Fe$_2$N-Type
(Nb$_2$C; ϵ-V$_2$C)

ζ-Fe$_2$N-Type
(V$_2$C; Mo$_2$C)

ξ-Nb$_2$C

Fig. 6. A view perpendicular to the carbon layers shown in Fig. 5 shows that ordering of the carbon atoms results in an overall symmetry of either the hexagonal or orthorhombic type. In the legend, the notations $C = 0$ and $C = \frac{1}{2}$ designate the two different positions of the carbon layers along the C axis of the parent L_3' structure. ⊗ : carbon random on $\frac{1}{2}$ the sites; O: carbon in the layer at $C = \frac{1}{2}$; ⊘: carbon in the layer at $C = 0$. [After Yvon et al. (12).]

between structures is expected to be small and possibly influenced by impurities.

Yvon et al. (12) and Rudy et al. (13) have reviewed the ordering in the Me$_2$C subcarbides of the fifth and sixth groups. In the Me$_2$C subcarbides,

Fig. 7. Position of vacancies (little boxes) in the crystal structure of V$_8$C$_7$. In this structure there is a helicoil arrangement of vacancies. [After Froidevaux and Rossier, J. Phys. Chem. Solids **28**, 1197 (1967). Reprinted by permission of Pergamon Press.]

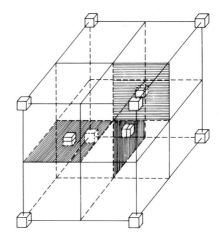

one half of the octahedral interstitial sites are vacant. On the basis of X-ray analysis, these phases had previously been classified as belonging to the L_3' structure with carbon being statistically distributed in one-half of the octahedral sites. The L_3' structure with the statistical distribution of carbon is apparently the stable high temperature form of V_2C, Nb_2C, Ta_2C, Mo_2C, and W_2C. The low temperature modifications of these phases have carbon ordered as listed in Table VI. Figure 5 shows the type of ordering that occurs in these low temperature modifications. In many cases, ordering of the carbon atoms changes the overall symmetry of the structure from hexagonal to orthorhombic (Fig. 6).

Ordering of carbon atoms has also been observed in nonstoichiometric monocarbides. Neutron diffraction studies on TiC_{1-x} by Gorbunov (14) have

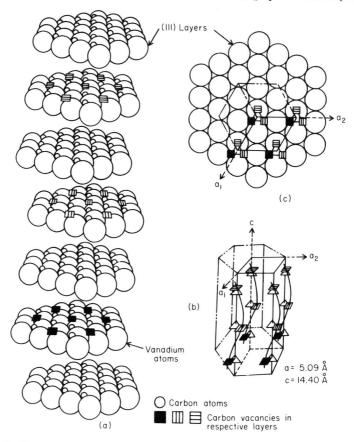

Fig. 8. The structure of V_6C_5 showing: (a) location of ordered carbon atom vacancies (*c*-axis expanded for clarity), (b) hexagonal superlattice formed by vacancy ordering, and (c) projection of superlattice on a {111} plane of vanadium atoms. Symmetry of superlattice corresponds to trigonal class $P3_1$ or its enantiomorph $P3_2$. The material sudied, $VC_{0.84}$, exhibits this structure, but contains a slight excess of carbon atoms over the integral composition. [From Venables *et al.* (20).]

shown that over most of the composition range carbon is random on the interstitial sites. Goretzki (15) has shown, however, carbon ordering in the composition range 32–36 at. % C ($\sim Ti_2C$). The new cell belongs to the O_h^7-$Fd3m$ space group with carbon occupying the octahedral sites at $16(c)$ between the metal atoms in $32(e)$ positions. The unit cell is about twice the size of the $B1$ subcell ($a = 8.6$ Å). The fractional coordinate x of the metal atoms is 0.245 which indicates that the metal atoms are shifted toward the carbon atoms. In the undistorted case x would equal 0.250. This shift also enlarges the unoccupied octahedral interstitial sites; in other words, the Ti atoms move away from the carbon vacancies. The shift could indicate strong Ti–C bonding or the presence of a charge associated with the vacant site.

The nuclear magnetic resonance and X-ray studies on VC_{1-x} at the

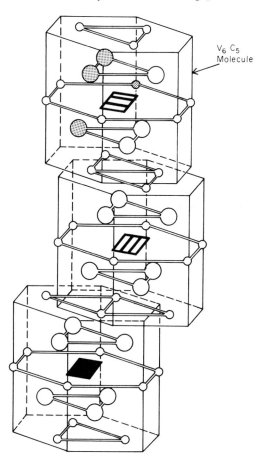

$V_6 C_5$
Molecule

Fig. 9. The structure of V_6C_5 viewed in terms of spiralling chains of V_6C_5 molecules. At the center of each molecule is a vacant carbon site (□) which corresponds to a vacancy on one of the 3–fold screw axes shown in Fig. 8b. A complete unit cell consists of nine such molecules, and, thus, contains 54 vanadium atoms (●) and 45 carbon atoms (●). [From Venables et al. (20).]

composition V_8C_7 indicate carbon ordering where again the vanadium atoms move away from the carbon vacancies (16–19). The proposed crystal structure for V_8C_7 (Fig. 7) also suggests that there is a mutual repulsion between carbon vacancies since the minimal distance between two vacancies equals that between third neighbors. Strain energy considerations would suggest that the vacancies should tend to cluster, and therefore the repulsive interactions of vacancies in V_8C_7 indicate that each vacancy has associated with it an electric charge.

Venables et al. (20) have used electron diffraction analysis and nuclear magnetic resonance to study the ordering in V_6C_5. This phase has hexagonal symmetry as contrasted to the cubic symmetry of V_8C_7. The proposed crystal structure for V_6C_5 is shown in Figs. 8 and 9. The complex ordering scheme of carbon vacancies again indicates a long-range interaction.

Other as yet unidentified phases have been observed in several carbide phase diagrams. These phases also probably exhibit ordering of carbon atoms. These phases will be discussed in Chapter 3.

III. Non-Hägg-like Phases

It is often supposed that the crystal structures of binary carbides and nitrides are all of the Hägg type; that is, close-packed or hexagonal arrays of metal atoms with carbon or nitrogen occupying the octahedral or trigonal prism interstitial sites. Recent neutron and electron diffraction studies, however, have shown the existence of several complex binary carbides and nitrides which cannot be described as Hägg phases. Many of these phases have been prepared as thin films and therefore may not be characteristic of the bulk samples. Others such as ϵ-Ti_2N and Cr_3C_2 are definitely characteristic of the bulk.

IIIA. ϵ-Ti_2N Phase

ϵ-Ti_2N(C4) has an unusual tetragonal structure which is related both to the bcc β-Ti and to the hcp α-Ti. Holmberg (21) has determined atomic positions using single crystal X-ray diffraction. The atomic positions and space group are as follows:

Space group: $P4_2/mnm$
Unit cell dimensions: $a = 4.9452$ Å, $c = 3.0342$ Å
Unit cell content: $2Ti_2N$

Atomic positions: 4Ti in 4(f): $x,x,0$; $\bar{x},\bar{x},0$; $\frac{1}{2}+x$, $\frac{1}{2}-x$, $\frac{1}{2}$; $\frac{1}{2}-x$, $\frac{1}{2}+x$, $\frac{1}{2}$;
$x_{Ti} = 0.296$
2N in 2(a): $0,0,0$; $\frac{1}{2},\frac{1}{2},\frac{1}{2}$

Figure 10a shows the projection of the structure parallel to [001] and illustrates the relationship of ϵ-Ti$_2$N to the bcc β-Ti. Figure 10b shows the projection parallel to [010] and illustrates the relation to the hcp α-Ti. It can be seen in both figures that the presence of N in the structure distorts the normal bcc and hcp positions.

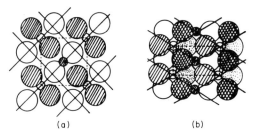

(a)   (b)

Fig. 10. (a) Shows the [001] projection of the ϵ-Ti$_2$N crystal structure and illustrates the relationship of this structure to bcc Ti. The large and small circles represent Ti and N atoms. The light circles have $z = 0$ and the dark ones $z = \frac{1}{2}$. The full lines correspond to the bcc Ti unit cell and show how the bcc structure is distorted in ϵ-Ti$_2$N. (b) Shows the [010] projection and illustrates its relationship to hcp Ti. The Ti positions are $y = 0.204$, 0.296, 0.704, and 0.796, as indicated by circles of increasing darkness. The N positions are $y = 0$ for the light small circles and $y = \frac{1}{2}$ for the dark circles. [After Holmberg (21).]

The nitrogen atoms are in the octahedral interstitial sites. The titanium atoms form octahedra that are joined by edges to form strings running parallel to the c-axis. Neighboring strings of octahedra are joined by sharing corners; Fig. 11 shows a layer of these octahedra.

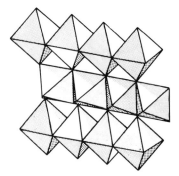

Fig. 11. In ϵ-Ti$_2$N, the Ti atoms are at the corners of an octahedron and N is in the center of the octahedron. The octahedra form layers by sharing edges and corners with neighboring octahedra.

The octahedral layer shown in Fig. 11 is also related to the octahedral layers found in the Nowotny subcell structures that will be discussed later in this chapter.

IIIB. ε-TaN Phase

The crystal structure of ε-TaN($B35$) is hexagonal with $a = 5.1808\ kX$, $c = 2.9049\ kX$ (22, 23). Tantalum occupies the $0\ 0\ 0$, $\frac{1}{3}\frac{2}{3}\frac{1}{2}$, $\frac{2}{3}\frac{1}{3}\frac{1}{2}$, and nitrogen the $\frac{1}{2}\ 0\ 0$, $0\ \frac{1}{2}\ 0$, $\frac{1}{2}\frac{1}{2}\ 0$ positions in the $P6/mmm$ (or $P62$) space group.

The crystal structure is unusual because the tantalum atoms form alternating layers of centered and uncentered hexagons. The N atoms are in the centers of deformed octahedra. These layers are shown in Fig. 12.

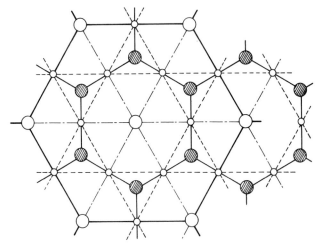

Fig. 12. Hexagonal networks in the crystal structure of ε-TaN. ⊘ : Ta atoms in $\frac{1}{3}\frac{2}{3}\frac{1}{2}$ and $\frac{2}{3}\frac{1}{3}\frac{1}{2}$; ◯ : Ta atoms in $0\ 0\ 0$; ○ : N atoms in $\frac{1}{2}\ 0\ 0$, $0\ \frac{1}{2}\ 0$, and $\frac{1}{2}\frac{1}{2}\ 0$.

The Ta atoms in this structure are very closely packed; the interatomic distances between Ta atoms being comparable to those in bcc Ta. Every atom at the atomic position $0\ 0\ 0$ has 14 neighboring Ta atoms, while those at $\frac{2}{3}\frac{1}{3}\frac{1}{2}$, $\frac{1}{3}\frac{2}{3}\frac{1}{2}$ have 11 neighboring Ta atoms.

IIIC. W–N Layered Structures

Pinsker and co-workers (24) have discovered an entire class of tungsten nitrides that do not belong to the Hägg type of structures. While these phases were prepared as thin films and probably do not represent bulk structures, the W–N phases are important for a number of reasons. First, they represent the largest group of non-Hägg-like compounds, second, they demonstrate new principles for the crystallography of nitrides, and third they illustrate a direct relationship to other types of interstitial compounds, particularly borides. Table VII lists the crystal structures and atomic positions of the tungsten nitrides.

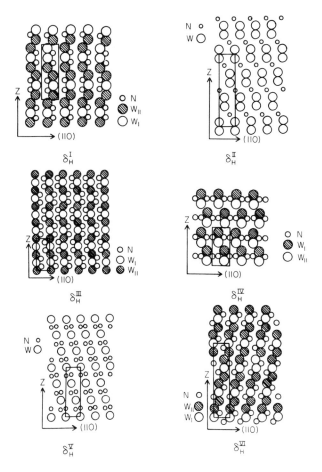

Fig. 13. Crystal structures of tungsten nitrides prepared as thin films. The subscripts "H" and "R" denote hexagonal and (ortho)rhombic, respectively. The large open circles denoting W_I mean that this W layer is not defective. The large hatched circles denoting W_{II} mean that this layer is defective with a random occupancy of the W sites. Projections are on (110) planes. [After Khitrova and Pinsker (24).]

TABLE VII. CRYSTAL STRUCTURES OF TUNGSTEN NITRIDES PREPARED AS THIN FILMS[a]

Symbol	Lattice constants, A	Space group	Position of W		N	Rough formula and theoretical density, g/cm³	Minimal distances, A	
			undefective	defective			W – N	N – N
1	2	3	4	5	6	7	8	9
δ_H^I	$a=2.885$, $c=15.30$—15.46, $c/a=5.3$—5.35	$P6_3/mmc$—D_{6h}^4	W_I at 2 (c): $1/3\ 2/3\ 1/4$, $2/3\ 1/3\ 3/4$	W_{II} at 4 (f): $1/3\ 2/3\ z$, $2/3\ 1/3\ 1/2\ +\ z$, $1/3\ 2/3\ 1/2-z$, $z=0,06$	N at 4 (f): $z=-0,165$	$W_{1.15}N$ — $W_{1.35}N$; $\rho=13.6$—15.7	W_I — N 2.16; W_{II} — N 2.26	2.60—2.62
δ_H^{II}	$a=2.89$, $c=22.85$, $c/a=7.9$	$P\bar{3}$—C_{3i}^1	W_I at 2 (c): $00\bar{z}$, $z_I=0.0607$; W_{II} at 2 (d): $1/3\ 2/3\ z$, $z_2=1/3-z_I$; W_{III} at 2 (d): $z_3=1/3+z_I$		N at 1 (b): $001/2$ N_{II} at 2 (d) $z=0.154$	W_2N; $\rho=12,0$	W_I — N 2.88; W_{II} — N 3.09; W_{III} — N 2.91	2.89
δ_H^{III}	$a=2.87$, $c=11.00$, $c/a=3.81$	$P6_3$—C_6^6	W_I at 2 (b): $001/4$, $003/4$	W_{II} at 2 (a): 000 $001/2$	N at 4 (f): $z=0.126$	$W_{0.64}N$; $\rho=11.0$	W_I — N 2.14; W — N 2.16	2 75
δ_H^{IV}	$a=2.89$, $c=10.8$, $c/a=3.73$	$P6_3$—C_6^6	W at 2 (b): $1/3\ 2/3\ z$, $2/3\ 1/3\ 1/2+z$ $z=0.625$	W_{II} at 2 (b): $z=0.375$	N_I at 2 (a): $00z$, $001/2+z$, $z=0.25$ N_{II} at 2 (b) $z=0.25$	$W_{0.6}N$; $\rho=10.63$	W_I — N 2.16; W — N	1.67
δ_R^V	$a=2.89$, $c=16.4$, $c/a=5.67$	$R\bar{3}m$—D_{3d}^5	W at 3 (a): 000, $1/3\ 2/3\ 2/3$, $2/3\ 1/3\ 1/3$		N at 6 (c): $(000,\ 1/3\ 2/3\ 2/3,\ 2/3\ 1/3\ 1/3)+00z$, $00z$ $z=0,1785$	$W_{0.5}N$; $\rho=9.0$	W — N: 2.915, 3.03, 3.04	1.71
δ_R^{VI}	$a=2.89$, $c=23.35$, $c/a=8.07$	$R\bar{3}m$—D_{3d}^5	W_I at 3 (a): $(000,\ 1/3\ 2/3\ 2/3,\ 2/3\ 1/3\ 1/3)$	W_{II} at 6 (c): $z=0.120$	N at 6 (c): $z=0.277$	$W_{1.17}N$; $\rho=13.6$	W_I — N 2.13; W_{II} — N 2.12	3.12

[a] After Khitrova and Pinsker (24).

The tungsten nitrides are built up from W and N atomic layers. Untypically, however, the W layers can be highly defective. Some of the N layers are also irregular, In δ_H^{IV} and δ_R^{V} the N atoms form uncentered hexagons; each atom is shared by three hexagons. In δ_R^{V} the N layers are slightly buckled. The interatomic distance between neighboring N atoms is only 1.67 Å in δ_H^{IV} or only about 10% greater than the usual atomic diameter for N. These short N–N distances signify probable covalent bonds. These structures are the first known among nitrides in which the N–N distance is small enough to indicate covalent bonds.

The graphite-like N layers in δ_H^{IV} and δ_R^{V} are also interesting because they show a relation between the nitrides and borides. The N layers in δ_H^{IV} and δ_R^{V} are exactly the same as those formed by boron atoms in AlB_2, W_2B_5, and $\epsilon\text{-}Mo_2B_5$. This link is important because many research articles have suggested that borides were distinct in their crystal structures and other properties from carbides and nitrides.

Figure 13 illustrates the layered structure of the tungsten-nitrides. The large cross-hatched circles denote the W positions taken up randomly in defective W layers.

IIID. Cr_2C_3 Structure

Another structure in which the metal atoms are no longer close-packed or nearly close-packed is $Cr_3C_2(D5_{10})$. Atomic positions have been determined with neutron diffraction techniques by Meinhardt and Krisement (25) following earlier work by Westgren (26). The unit cell is orthorhombic with

$$a = 11.46 \text{ Å}, \quad b = 5.52 \text{ Å}, \quad \text{and} \quad c = 2.821 \text{ Å}.$$

The space group is $D_{2h}^{16}\text{-}Pbnm$ with atomic positions at

$$(x, y, \tfrac{1}{4}); \quad (\tfrac{1}{2}-x, \tfrac{1}{2}+y, \tfrac{1}{4}); \quad (\bar{x}, \bar{y}, \tfrac{3}{4}); \quad \text{and} \quad (\tfrac{1}{2}+x, \tfrac{1}{2}-y, \tfrac{3}{4}).$$

Cr and C atoms are located at the following positions:

$4Cr_I$	in (c) at $x =$	0.406,	$y =$	0.030	
$4Cr_{II}$	in (c) at $x =$	$-0.230,$	$y =$	0.175	
$4Cr_{III}$	in (c) at $x =$	$-0.070,$	$y =$	-0.150	
$4C_I$	in (c) at $x =$	0.204,	$y =$	0.092	
$4C_{II}$	in (c) at $x =$	$-0.048,$	$y =$	0.228	

The Cr atoms form trigonal prisms which join other trigonal prisms at common edges. The C atoms are at the centers of the trigonal prism (see Fig. 14).

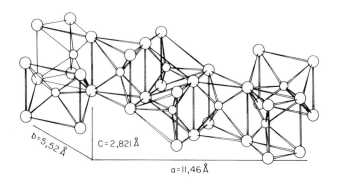

Fig. 14. In the crystal structure of Cr_3C_2 the Cr (large circles) atoms are at the corners of trigonal prisms and the carbon (small circles) is in the center of the prism. [After Meinhardt and Krisement (*25*).]

IV. Nowotny Octahedral Phases

In the Hägg phases, the metal atoms form a close-packed or nearly close-packed structure and the nonmetal atoms occupy the available interstitial sites. Nowotny and co-workers, particularly Jeitschko and Benesovsky (*27*, *28*) have discovered nearly 200 ternary compounds involving refractory transition elements, metalloids, and carbon or nitrogen, *in which the metal substructure is no longer close-packed or nearly close-packed*, but in which the nonmetal atom occupies an octahedral interstitial site. The crystal structures of these ternary phases are best described in terms of the geometric arrangements of these octahedra. The transition-metal atoms are at the corners of an octahedron, and the nonmetal is at the center of the octahedron. These octahedra are generally not ionic, and, therefore, not all the interstitial sites need be occupied as in the case of the SiO_4 and $Al(OH)_6$ coordination polyhedra in the silicates and aluminides. For convenience, we shall abbreviate the octahedral unit by $[Me_6X]$ even though the interstitial site need not be occupied in these nonionic octahedra. Here Me atoms are generally restricted to the III(Sc, Y, La), IV(Ti, Zr, Hf), V(V, Nb, Ta), and VI(Cr, Mo, W) group transition metals.

The Nowotny octahedral phases have the general formula $Me_aM_bX_c$ and include the following types: interstitial β-manganese type Me_3M_2X, the *H*-phases Me_2MX, the Perovskite type Me_3MX, and the η-carbide structure type ($E9_3$), Me_4M_2X. In these formulas, M generally refers to a nontransition metal such as Zn, Cd, Al, Ga, In, Tl, Sn, and Pb, or to Ge and Si. For the Perovskite phases, Me also includes Mn, Fe, Co, Ni, Pd, and Pt.

It is interesting that the Nowotny octahedral phases generally involve three elements with widely different melting points. The M elements usually have very low melting points. The phases are, therefore, difficult to prepare, and for this reason their existence has only recently been discovered.

In all these phases, carbon or nitrogen atoms are always in an octahedral interstitial site composed only of Me atoms. In the case of the β-Mn type, the octahedra are slightly distorted.

The discovery of these phases by Nowotny and co-workers has revealed the importance of the $[Me_6X]$ octahedron as a structural unit. The $[Me_6X]$ octahedron exists, of course, in the NaCl and L_3' crystal structures, but the description of these phases in terms of coordination polyhedra is somewhat cumbersome. In Nowotny octahedral phases, octahedra are arranged in more complex patterns than in NaCl or L_3'. The octahedra can share common corners, edges, or faces and in so doing form one, two, or three-dimensional structures. Alternating the manner of contact between individual $[Me_6X]$ octahedra alters the Me : X ratio and also the Me : M ratio.

Not all the Nowotny octahedral phases are refractory compounds. Nevertheless these compounds are included in this discussion because the principles involved in forming these crystal structure types increase our understanding of the crystal chemistry of the refractory carbides and nitrides.

IVA. Crystal Structures in Which Two Octahedra Share Corners

(1) Me_3M_2X Interstitial β-Mn type. In the β-Mn type, $[Me_6X]$, octahedra are joined by sharing corners. Since each Me atom is shared by two octahedra, the Me : X ratio is three—provided, of course, that all inter-

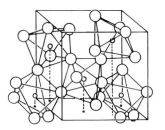

Fig. 15. Filled β-Mn type structure of cubic Me_3M_2X, showing octahedra sharing corners. Only Me and X atoms are shown. ◯ : Me atoms forming distorted octahedra ; ○ : interstitial C or N. [After Nowotny and Benesovsky, *in* "Phase Stability in Metals and Alloys" (P. S. Rudman, J. Stringer, and R. I. Jaffee, eds.) copyright © 1967 by McGraw-Hill, Inc. Used with permission of McGraw-Hill Book Company.]

stitial sites are filled. The octahedra in this structure are slightly distorted as shown in Fig. 15. The positions of the M atoms are not shown for sake of clarity. They form, however, a rather open structure. Table VIII lists the β-Mn phases according to the group number of the M element.

TABLE VIII

FILLED β-Mn TYPE, CUBIC COMPOUNDS
AND LATTICE PARAMETERS[a]

Compound	Lattice parameter (Å)
Nb_3Al_2C	7.07_2–7.07_9[b]
$Nb_3(Au_{2/3}Ga_{1/3})_2C_x$	7.084
Ta_3Al_2C	7.03_8
Mo_3Al_2C	6.86_0–6.86_6[b]
V_3Zn_2N	6.60_6
V_3Ga_2N	6.62_0
$V_3(Au_{2/3}Ga_{1/3})_2N_x$	6.726
Nb_3Al_2N	7.03_4
$Nb_3Au_2N_x$	7.085
$Nb_3(Au_{2/3}Zn_{1/3})_2N_x$	7.052
$Nb_3(Au_{2/3}Ga_{1/3})_2N_x$	7.048
$Ta_3(Au_{2/3}Ga_{1/3})_2N_x$	7.034

[a] After Nowotny and Benesovsky, *in* "Phase
Stability in Metals and Alloys" (P. S. Rudman,
J. Stringer, and R. I. Jaffee, eds.) copyright
© 1967 by McGraw-Hill, Inc. Used with
permission of McGraw-Hill Book Company.
[b] Al deficient.

(2) Me₃MX—Perovskite carbides and nitrides (CaTiO₃ type, E2₁). In
the Perovskite structure the [Me₆X] octahedra are linked by sharing corners
and form a simple cubic array as shown in Figs. 16a and 16b. In this

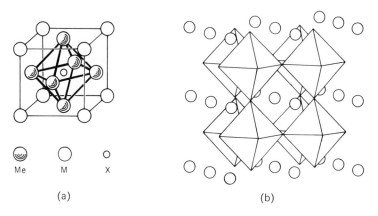

Me   M   X

(a)   (b)

Fig. 16. (a) Unit cell of the Perovskite structure showing the [Me₆X] octahedra surrounded by M atoms arranged in a simple cubic structure. (b) Arrangement of [Me₆X]
octahedra and M atoms in four-unit cells of Perovskite structure. Octahedra share
corners. [Part (a) is after Nowotny *et al.* (*28*).]

TABLE IX. PEROVSKITE CARBIDES AND NITRIDES BY GROUP NUMBER OF THE TRANSITION-METAL ELEMENT[a]

III	IV	V	VI	VII	VIII (Fe)	VIII (Co)	VIII (Ni)
Sc_3AlC	Ti_3AlC Ti_3InC Ti_3TlC Ti_3InN Ti_3TlN $Ti_3Au(N,O)_{1-x}$	$V_3Au(N,O)_{1-x}$	—	Mn_3ZnC Mn_3AlC Mn_3GaC Mn_3InC Mn_3GeC Mn_3SnC Mn_3ZnN[b]	Fe_3ZnC Fe_3AlC Fe_3GaC Fe_3InC Fe_3GeC Fe_3SnC Fe_3MgN[b] Fe_3ZnN[b] Fe_3AlN[b] Fe_3GaN[b] Fe_3InN[b] Fe_3GeN[b] Fe_3SnN[b]	Co_3MgC Co_3ZnC Co_3AlC Co_3GaC Co_3InC Co_3GeC Co_3SnC Co_3ZnN Co_3GaN Co_3InN Co_3GeN Co_3SnN	Ni_3MgC Ni_3ZnC Ni_3AlC[b] Ni_3GaC[b] Ni_3InC Ni_3GeC[b] Ni_3ZnN Ni_3AlN Ni_3InN
Y_3AlC Y_3TlC_{1-x}	—	—	—	—	—		Pd_3AlC[b] Pd_3InC Pd_3PbC
La_3InC_{1-x}	—	—	—	—	—		Pt_3MgC Pt_3ZnC[b] Pt_3HgC[b] Pt_3AlC Pt_3InC Pt_3SnC Pt_3PbC

Rare Earth Perovskite Phases

III	IV	V	VI	VII	VIII (Fe)	VIII (Co)	VIII (Ni)
La_3InC_{1-x}	Ce_3InC_{1-x} Ce_3TlC_{1-x} Ce_3SnC_{1-x} Ce_3PbC_{1-x}	Pr_3GaC_{1-x} Pr_3InC_{1-x} Pr_3TlC_{1-x} Pr_3SnC_{1-x} Pr_3PbC_{1-x}	$Nd_3AlC_{0.9}$	...	$Gd_3AlC_{0.9}$ ...	$Dy_3AlC_{0.7}$ Dy_3GaC_{1-x} Dy_3InC_{1-x} Dy_3SnC_{1-x}	$Ho_3AlC_{0.7}$

[a] Also included are rare-earth Perovskite carbides and nitrides (see 27–31).

[b] Extending out from the binary phase which has the Cu_3Au-type structure.

structure, the M atoms are also in a cubic pattern around the octahedra. There are a great many compounds with this structure, as listed in Table IX (27–31). The surprising feature about this structure is the wide selection of transition elements originating from different parts of the periodic table. The transition elements are usually Mn, Fe, Co, Ni, Pd, and Pt, but examples have been found where the transition elements belong to earlier groups. The Me element can also be a rare earth element (27). Most of these phases with Me members of the seventh or eighth groups have been found by Stadelmaier and co-workers (31).

Referring to Table II which lists the radius ratios of the transition elements with C or N, we find that, for the carbides of Mn, Fe, Co, and Ni, the r_C/r_{Me} ratio is greater than 0.59, which is the limit for octahedral interstitial occupancy according to the Hägg rule. It is therefore necessary to extend the upper limits of the Hägg rule to include these structures (31).

(3) κ-Carbides and κ-Like Complex Carbides. The κ-phases are another

TABLE X

LATTICE PARAMETERS OF κ-CARBIDES AND
κ-LIKE COMPLEX CARBIDES[a, b]

System	Parameter		Ratio
	a (Å)	c (Å)	c/a
$W_{10}Co_3C_4$	7.848	7.848	1.000
$W_9Co_3C_4$	7.826	7.826	1.000
$W_{16}Ni_3C_6$	7.818	7.818	1.000
W—Ni—C	7.848	7.848	1.000
Mo—Mn—Al—C	7.87_6	7.86_7	0.999[c]
Mo—Fe—Al—C	7.849	7.848	1.000
Mo—Co—Al—C	7.95_0	7.84_3	0.986
Mo—Ni—Al—C	7.89_3	7.85_0	0.995
$Mo_{12}Cu_3Al_{11}C_6$	7.95_2	7.865	0.989
W—Mn—C	7.756	7.756	1.000
W—Mn—Al—C	7.90_3	7.787	0.986
W—Fe—Al—C	7.89_5	7.85_7	0.995
Mn_3Al_{10}	7.543	7.898	1.048
Mn_3Al_9Si	7.513	7.745	1.031
Co_2Al_5 ($D8_{11}$ type)	7.656	7.543	0.992

[a] After Nowotny and Benesovsky, in "Phase Stability in Metals and Alloys" (P. S. Rudman, J. Stringer, and R. I. Jaffee, eds.) copyright © 1967 by McGraw-Hill, Inc. Used with permission of McGraw-Hill Book Company.
[b] Also included are the intermetallic compounds (without nonmetal) representing essentially the parent lattice.
[c] Homogeneous range.

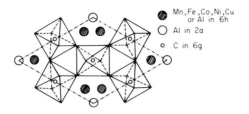

Mn, Fe, Co, Ni, Cu
or Al in 6h

Al in 2a

C in 6g

Fig. 17. In the κ (kappa)-type carbide structure the [Me$_6$X] are joined at the corners Nontransition metal-atoms such as Al can be accommodated in the 2a positions [After Nowotny and Benesovsky, *in* "Phase Stability in Metals and Alloys" (P. S. Rudman, J. Stringer, and R. I. Jaffee, eds.) copyright © 1967 by McGraw-Hill, Inc. Used with permission of McGraw-Hill Book Company.]

example of a structure in which the [Me$_6$X] octahedra are joined by common corners to form a three-dimensional network with three- and six-fold channels. Rautala and Norton (*32*) found the first κ-phase, $W_{10}Co_3C_4$. Its crystal structure and correct composition, $W_9Co_3C_4$, were subsequently elucidated by Schönberg (*33*). The structure has hexagonal symmetry with the c axis very nearly equal to a axis and belongs to the D_{6h}^4-$P6_3/mmc$ space group. Atomic positions are as follows:

$3Co + 3W$ in $6(h)$–x, $2x$, $\frac{1}{4}$ with $x = 0.890(I)$
$3Co + 3W$ in $6(h)$–x, $2x$, $\frac{1}{4}$ with $x = 0.550(II)$
$12W$ in $12(k)$–x, $2x$, z with $x = 0.205$ and $z = 0.075$
$2C$ in $2(c)$–$\frac{1}{3}$, $\frac{2}{3}$, $\frac{1}{4}$
$6C$ in $6(g)$–$\frac{1}{2}$, 0, 0

The metal atoms in the $6(h)$II and $12(k)$ positions form a network of slightly deformed octahedra in which all corner atoms are common to two such octahedra. The carbon atoms in the $6(g)$ sites are located in the centers of these octahedra.

Kuo (*34*) found indications for the existence of corresponding phases in the W–{Mn, Ni}–C systems, and Whitehead and Brownlee (*35*) reported a κ-phase with the composition $W_{16}Ni_3C_6$. Nowotny and Benesovsky (*27*) reported that Al is a suitable element for stabilizing these complex phases (see Table X). Al substitutes for the transition elements and also occupies the $2(a)$ positions, as illustrated in Fig. 17. By filling the $2(a)$ positions, the parent structure of these κ-carbides is made to resemble that of Mn_3Al_{10} or Mn_3Al_9Si. Nowotny and Benesovsky (*27*) further suggest that in the Al-stabilized phases, carbon occupies only the $6(g)$ octahedral sites because the Me—C bonds are stronger than Al—C bonds.

IVB. Crystal Structure in Which the Octahedra Share Common Edges

(1) H Phases or Cr₂AlC Type (Me₂MX). The *H*-phases have a very simple structure which is closely related to the L_3' type. In this structure, the [Me₆X] octahedra are arranged in layers as shown in Fig. 18. In this

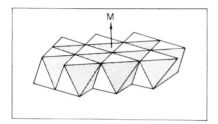

Fig. 18. [Me₆X] octahedra arranged in a layer (type I) formed by octahedra sharing edges. In the H phases the M atoms are located directly above the triangular-shaped voids. (See arrow.)

planar structure, each [Me₆X] octahedron shares six of its twelve edges with adjacent octahedra. Each Me atom is shared by three octahedra. Thus, the chemical formula for the planar structure is Me₂X. The layer has hexagonal symmetry with the sequential ordering of Me atomic layers being AB. Figure 18 also illustrates tetrahedral voids created by the joining of the octahedra. The Cr₂AlC structure is constructed by placing M atoms over the tetrahedral voids and placing another planar layer of octahedra over the first so that the voids line up one over another.

The Cr₂AlC structure is shown in Fig. 19. The sequential ordering of atomic layers is —MAXBMBXAM—. Phases belonging to this group are shown in Table XI. The Cr₂AlC type is closely related to the γ′-MoC type

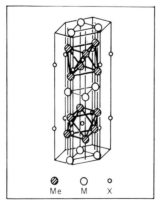

Me   M   X

Fig. 19. Crystal structure of the *H*-phase Me₂MX showing the individual octahedra that form the layers of octahedra shown in Fig. 18. [After Nowotny *et al.* (*28*).]

TABLE XI

H PHASES AND LATTICE PARAMETERS[a]

H phase	*a* (Å)	*c* (Å)	*H* phase	*a* (Å)	*c* (Å)
Ti_2CdC	3.09_9	14.41_4	V_2AlC	2.91_3	13.1_4
Ti_2AlC	3.04	13.6_0	V_2GaC	2.93_8	12.8_4
Ti_2GaC	3.06_4	13.30_5	V_2GaN	2.89_2	12.93
Ti_2InC	3.13_2	14.06	V_2GeC	3.00_1	12.25
Ti_2AlN	2.99_4	13.61			
Ti_2GaN	3.00_4	13.30_2	Nb_2AlC	3.10_3	13.8_3
Ti_2InN	3.07_4	13.97_6	Nb_2GaC	3.13_1	13.56_5
Ti_2GeC	3.07_9	12.9_3	Nb_2InC	3.17_2	14.37
Ti_2SnC	3.18_6	13.63	Nb_2SnC	3.24_5	13.77
Zr_2InC	3.34_7	14.90	Ta_2AlC	3.07_5	13.8_3
Zr_2TlC	3.36_3	14.78_9	Ta_2GaC	3.10_4	13.57
Zr_2InN	3.27_7	14.83_6			
Zr_2TlN	3.30_1	14.71	Cr_2AlC	2.86_6	12.8_2
Zr_2SnC	3.34_7	14.5_9	Cr_2GaC	2.88_6	12.61_6
Zr_2PbC	3.38_4	14.66_6	Cr_2GeC	2.95_4	12.08
Hf_2InC	3.30_7	14.73	Mo_2GaC	3.01_7	13.18
Hf_2TlC	3.32_2	14.62_5			
Hf_2InN	3.23_0	14.74			
Hf_2SnN	3.31_2	14.3_9			
Hf_2PbC	3.35_8	14.46_5			

[a] After Nowotny and Benesovsky, *in* "Phase Stability in Metals and Alloys" (P. S. Rudman, J. Stringer, and R. I. Jaffee, eds.) copyright © 1967 by McGraw-Hill, Inc. Used with permission of McGraw-Hill Book Company.

(TiP), L_3' and NaCl structures. γ'-MoC type is obtained by replacing M atoms with carbon atoms. The sequential ordering is thus —AXBXBXA—. The L_3' structure is derived from the Cr_2AlC type by removing the M layers so that the octahedral planar structures join each other along top and bottom planes. This results in a sequential ordering of atomic layers of —AXBXAXBX—.

(2) Mo_2BC. A second example of a structure in which octahedra are joined through sharing edges is the Mo_2BC structure. This phase is somewhat different from the other Nowotny octahedral phases by having boron substituted for the M metal. The phase is also unusual because it is the only known pure ternary compound involving a refractory transition-metal atom, boron, and carbon (*36*). There are, however, several boron–carbon compounds with thorium and uranium (*37*). Jeitschko *et al.* (*38*) found that the crystal structure of Mo_2BC is an interesting combination of boride and

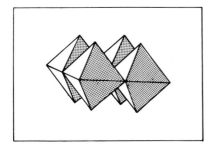

Fig. 20. Arrangement of octahedra sharing common edges to form a layer of octahedra (type II) as found in the crystal structure of Mo_2BC.

carbide subcells. The carbon atoms are centrally located in $[Mo_6C]$ octahedra which join edges to form a layer (Type II) as shown in Fig. 20. These Type II octahedral layers in Mo_2BC are in turn separated from one another by a boron layer (see Fig. 21). The boron atoms form zig-zag chains running through trigonal prisms of molybdenum atoms, as is typical of transition-metal monoborides (*1*). Jeitschko (*39*) further clarified the relationship of Mo_2BC to β-MoB, MoC(*B*1), and a similar orthorhombic structure MoAlB (see Fig. 21).

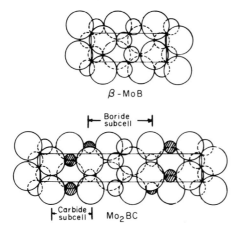

Fig. 21. The crystal structure of Mo_2BC is an interesting combination of alternating subcells of molybdenum boride and molybdenum carbide. The boride subcell is related to the crystal structure of β-MoB (top) and the carbide subcell is related to MoC with the *B*1 crystal structure. [After Jeitschko (*39*).]

The lattice parameters of the orthorhombic Mo_2BC are $a = 3.086$, $b = 17.35$, and $c = 3.047$ Å (38). The structure has the symmetry of the D_{2h}^{17} space group.

IVC. Crystal Structures in Which the Octahedra Share Common Faces

(1) η-Carbides and η-Like Carbides (E9₃). η-carbides have been studied extensively $(34, 40$–$55)$ primarily because of the great technical interest in the Co–W–C ternary system for the cutting tool industry. There are numerous η and η-like carbides. The η_1 carbide has the general formula Me_3M_3X, the η_2 carbide, Me_4M_2X. Here, Me is usually a group IV, V, or VI transition metal, and M is Cr, Mn, Fe, Co, Ni, or one of several other elements, depending on whether X is C, N, or O. In addition to the η_1 and η_2 carbides,

TABLE XII

ATOMIC POSITIONS FOR THE η_1 AND η_2 PHASES AT THE COMPOSITIONS
W_3Co_3C (η_1) AND W_4Co_2C (η_2) [a]

Positions in $O_h^7 - Fd3m$ [b]	η_1 W_3Co_3C	η_2 W_4Co_2C	Coordinate spacing for point positions
16 : (c)	16C	16C	—
16 : (d)	16Co	16W	—
32 : (e)	32Co	32Co	$x \sim 0.83$
48 : (f)	48W	48W	$x \sim 0.19$
8 : (a)	Sites for interstitial atoms in several η-like phases, but these sites are not occupied in η_1 or η_2.		

[a] Data are from Kiessling (43).
[b] There are two ways in which the η-carbide atomic positions can be described which differ by a translation of $\frac{1}{2}\frac{1}{2}\frac{1}{2}$. Both types appear in the literature. The notations of Mueller and Knott (53) differ from those given here in that the $16(d)$ become $16(c)$, the $8(a)$ become $8(b)$; for $32(e)$, one has

$$\tfrac{3}{4} - x_{Kiessling} = x_{Mueller\ and\ Knott,}$$

and for $48(f)$,

$$\tfrac{1}{2} - x_{Kiessling} = x_{Mueller\ and\ Knott.}$$

there are several η-like compounds which differ from the η_1 and η_2 phases primarily in the number of octahedral interstitial sites occupied by the interstitial atoms.

All η-carbides and η-like phases have an extended fcc crystal structure and have the symmetry of the $Fd3m$ space group. Westgren (40, 41) determined the atomic positions for the η_1 carbide W_3Co_3C; Kislyakova (42) and Kiessling (43) determined the positions for the η_2 carbide W_4Co_2C. These positions are given in Table XII.

The atoms in the 16(d) and 32(e) positions form a network of tetrahedra. The Me atoms in the 48(f) positions form a network of octahedra in which every second one is slightly distorted. The octahedra are joined together by shared faces. To visualize the structure, it is instructive first to imagine the arrangement of the regular octahedra and the tetrahedra. The large fcc cells of the η-carbide and the η-like phases can be broken down into eight cubic subcells of two alternating patterns, as shown in Fig. 22, which is

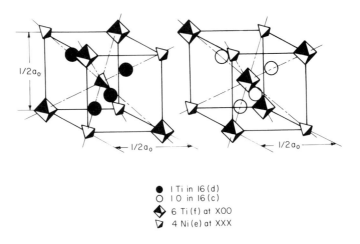

$\bullet$ I Ti in 16(d)
$\circ$ I O in 16(c)
$\blacklozenge$ 6 Ti(f) at X00
$\triangledown$ 4 Ni(e) at XXX

Fig. 22. Two different subcells of the phases in (a) and (b) show the arrangement of regular octahedra and tetrahedra. The octahedra and tetrahedra are shown collapsed about their centers for easier visualization. [After Mueller and Knott (53).]

based on the representation of Mueller and Knott (53). Their figure is based on neutron diffraction results on Ti_2Ni (the parent metal phase for the η-like phases) and Ti_4Ni_2O, an η-like phase. In Fig. 22, the regular octahedra and tetrahedra have been collapsed about their centers for easier visualization. The atomic positions in the figure have been changed to conform with those given in Table XII.

If we visualize the regular octahedra as a single atom (and neglect the other atoms) then there is a similarity between the structure shown in Fig. 22b and the diamond structure. The interstitial positions at the center of the regular octahedra are the $8(a)$ positions. For the η_1 and η_2 carbide phases, these sites are not occupied. The oxygen atoms in Fig. 22b are at the $16(c)$ positions. These are the positions which carbon occupies in the η_1 and η_2 phases. By allowing the collapsed regular octahedra to expand to their full size, slightly distorted octahedra of six Me atoms in $48(f)$ positions are formed about each oxygen of $16(c)$ site. Regular and distorted octahedra formed by the Me atoms at $48(f)$ sites are shown in Fig. 23 for the $a/2$ sized

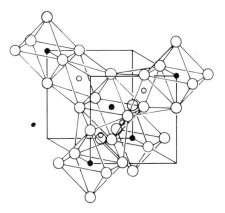

Fig. 23. By expanding the regular octahedra in Fig. 22b to their full size, slightly distorted octahedra are formed about the $16(c)$ interstitial sites. Each octahedron shares at least two faces with neighboring octahedra. Not all octahedra share the same number of faces. ○: Me atoms at $48(f)$; ●: Interstitial atoms at $8(a)$; ○: Interstitial atoms at $16(c)$. [Modified from a figure given by Nowotny and Benesovsky (27).]

subcell analogous to Fig. 22b. Here both the $16(c)$ and $8(a)$ interstitial sites are shown occupied.

By considering all the atomic positions $48(f)$, $32(e)$, $16(d)$, $16(c)$, and $8(a)$ Parthé et al. (55) have given a complete description of the different η-phases as shown in Table XIII. The phases have been grouped together according to the fraction of octahedral interstitial sites filled. Historically, the η-phases with two thirds of the interstitial sites filled have been subdivided into the η_1 and η_2 type phases.

Pearson (2, p. 925) has compiled a list of η_1 and η_2 carbides in which the M atoms are also transition elements; Nowotny and co-workers (27) have found numerous examples in which either all or part of the transition elements could be replaced by *meta*-metals such as Zn or Al or by the nontransition-metal elements Cu, Ag, and Au. These examples are listed in Table XIV.

(2) Filled D8₈ or Mn₅Si₃ Type. The interstitially filled $D8_8$ type of structure is characterized by parallel chains of $[Me_6X]$ octahedra within which each $[Me_6X]$ octahedron shares two opposite faces with neighboring

TABLE XIII. TYPES OF η AND η-LIKE PHASES $Fd3m$ (O_h^7) [a]

Ideal formula	Examples	Point positions (Kiessling) origin at 43m	Alternate point Group (Mueller and Knott)	Reference
$Me_6^I Me_2^{II} M_4 X_2 X_1$ (All interstitial sites filled)	$Nb_6Nb_2Zn_4C_2C_1$ $\equiv Nb_8Zn_4C_3$	Me^I in 48 (f), $x\sim0.19$; Me^{II} in 16 (d); M in 32 (e), $x\sim0.83$; X in 16 (c); X in 8 (a)	Me^I in 48 (f), $x\sim0.31$; Me^{II} in 16 (c); M in 32 (e), $x\sim0.91$; X in 16 (d); X in 8 (b)	(a)
$Me_6^I Me_2^{II} M_4 X_2$ ($\tfrac{2}{3}$ of interstitial sites filled)	$W_6W_2Co_4C_2$ $\equiv W_4Co_2C$ (η_2 type); $W_6Fe_2Fe_4C_2$ $\equiv W_3Fe_3C_2$ (η_1 type); $Ti_6Ti_2Ni_4O_2$ $\equiv Ti_4Ni_2O$	Me^I in 48 (f), $x\sim0.19$; Me^{II} in 16 (d); M in 32 (e), $x\sim0.83$; X in 16 (c)	Me^I in 48 (f), $x\sim0.31$; Me^{II} in 16 (c); M in 32 (e), $x\sim0.91$; X in 16 (d)	(b–f)
$Me_6^I Me_2^{II} M_4 X_1$ ($\tfrac{1}{3}$ of interstitial sites filled)	$W_6Fe_2Fe_4C$ $\equiv W_6Fe_6C$	Me^I in 48 (f), $x\sim0.19$; Me^{II} in 16 (d); M in 32 (e), $x\sim0.83$; X in 8 (a)	Me^I in 48 (f), $x\sim0.31$; Me^{II} in 16 (c); M in 32 (e), $x\sim0.91$; X in 8 (b)	(g)

[a] After Parthé et al. (55).

References for Table XIII

a. W. Jeitschko, H. Holleck, H. Nowotny, and F. Benesovsky, Monatsh. Chem. **95**, 1004 (1964).
b. A. Westgren, Jernkontorets Ann. **117**, 1 (1933).
c. E. N. Kislyakova, Zh. Fiz. Khim. **17**, 108 (1943).
d. R. Kiessling, 3rd Proc. Int. Symp. Reactiv. Solids (1952). p. 1065 (1954).
e. K. Kuo, Acta Met. **1**, 301 and 611 (1953).
f. M. H. Mueller and H. W. Knott, Trans. AIME **277**, 674 (1963).
g. J. Leciejewicz, J. Less-Common Metals **7**, 318 (1964).

TABLE XIV

η-LIKE PHASES IN WHICH M IS A NONTRANSITION-METAL ELEMENT[a]

Phase	Parameter (Å)	Phase	Parameter (Å)
$Ti_4Zn_2C_x$	11.55_8	$Ti_4Zn_2N_x$	11.49_6
$Zr_4Zn_2C_x$	12.16_0	$Zr_4Zn_2N_x$	12.13_1
$Hf_4Zn_2C_x$	12.04_4	$Hf_4Zn_2N_x$	11.96_6
$Nb_4Zn_2C_{x(x \sim 3/2)}$	11.74	$Nb_4Zn_2N_x$	11.54_7
Ta_3VAl_2C	11.6_7	$(Nb_3Zn_3N_x)$	
Nb_3CrAl_2C	11.7_1	Zr_3Cu_2Al	
Ta_3CrAl_2C	11.6_0	Zr_3Ag_2Al	
Nb_3MnAl_2C	11.6_9	Hf_3Cu_2Al	
Ta_3MnAl_2C	11.6_1	Hf_3Ag_2Al	
Nb_3FeAl_2C	11.6_3	Nb_3Cu_2Al	N- or O-containing
Ta_3FeAl_2C	11.5_7	$Zr—Cu—Zn$	
Ta_3CoAl_2C	11.5_6	$Zr—Ag—Zn$	
Nb_3Ni_2AlC	11.5_5	$Zr—Au—Zn$	
Ta_3Ni_2AlC	11.4_9	$Hf—Cu—Zn$	
Ta_3CuAl_2C	11.6_2	$Hf—Au—Zn$	

[a] After Nowotny and Benesovsky, *in* "Phase Stability in Metals and Alloys" (P. S. Rudman, J. Stringer, and R. I. Jaffee, eds.), copyright © 1967 by McGraw-Hill, Inc. Used with permission of McGraw-Hill Book Company.

octahedra (Fig. 24). This structure belongs to the D_{6h}^3-$P6_3/mcm$ space group. The atomic positions for the general chemical formula $Me_3^I Me_2^{II} M_3 X$ ($\equiv Me_5 M_3 X$) are listed below:

$$Me^I \text{ in } 6(g) \ x \sim 0.24$$
$$Me^{II} \text{ in } 4(d)$$
$$M \text{ in } 6(g) \ x \sim 0.61$$
$$X \text{ in } 2(b)$$

Here the x values refer to neutron diffraction results on $Mo_{4.8}Si_3C_{0.6}$ by Parthé *et al.* (*55*). The Me atoms in the $6(g)$ positions are those forming the one-dimensional alignment of octahedra shown in Fig. 24.

An important characteristic of the phases crystallizing in this structure is their broad composition range of stability. Generally, not all the interstitial sites at $2(b)$ are filled, and often only a few of these sites are occupied. There are numerous "interstitial free" binary $D8_8$-type phases $Me_5 M_3$ (*56*), numerous ternary solid solutions, $Me_5 M_3(X)$ extending out from the pure binaries, and a few "true" ternary compounds such as Zr_5Si_3C (*28*) and $Mo_{4.8}Si_3C_{0.6}$ (*57*). Table XV lists the phases with the filled $D8_8$ type of structure; the pure binaries are listed by Nowotny (*56*). Because the compounds do not form

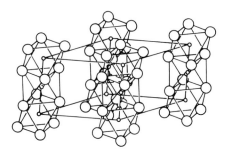

Fig. 24. [Me$_6$X] octahedra in the $D8_8$ structure are aligned parallel to the c-axis. Each octahedron shares two faces with adjacent octahedra. Generally not all the octahedral interstitial sites are filled. ◯ : Me atoms forming octahedra in 6(g) ; ○ : Interstitial C or N in 2(b). [After Nowotny and Benesovsky, *in* "Phase Stability in Metals and Alloys" P. S. Rudman, J. Stringer, and R. I. Jaffee, eds.) copyright © 1967 by McGraw-Hill, Inc. Used with permission of McGraw-Hill Book Company.]

at the ideal composition, a certain number of vacant sites and/or mixing of atoms in different sites is to be expected (*57*).

TABLE XV

INTERSTITIALLY FILLED $D8_8$ PHASES[a]

Phase	a (Å)	c (Å)	Reference
$Zr_5Si_3C_x$ (0.1–5 at. % C)	7.91$_4$–7.90$_9$	5.55$_9$–5.57$_9$	(a)
$Hf_5Ge_3C_x$ (3 at. % C)	7.88$_3$	5.53$_7$	(b)
$V_5Si_3(C)$	7.13$_5$	4.84$_2$	(c)
$Nb_5Ga_3C_x$ (5–10 at. % C)	7.72	5.27	(d)
$Ta_5Ga_3C_x$ (5–10 at. % C)	7.66$_1$	5.28$_0$	(d)
$Ta_{4.8}Si_3C_{0.5}$	7.49$_4$	5.242	(e)
$Ta_5Ge_3C_x$ (5–10 at. % C)	7.58$_6$	5.21$_7$	(d)
$Cr_{5-x}Si_{3-y}C_{x+y}$	6.99$_3$	4.72$_6$	(f)
$Mo_{4.8}Si_3C_{0.6}$	7.286	5.046	(g)
$W_{5-x}Si_{3-y}$ (C, N, O)$_{x+y}$	7.19	4.85	(f)
$Mn_5Si_3(C)$	6.91$_5$	4.82$_3$	(d)

[a] The following additional $D8_8$ phases have also been mentioned in the literature (h, i. j): $Zr_5Al_3(X)$, $Hf_5Al_3(X)$, $Hf_5Si_3(X)$, $Nb_5Si_3(X)$, $W_5Si_3(Cr, X)$, $Ti_5Ga_3(X)$, $V_5Ga_3(X)$, $V_5Ge_3(X)$, $Nb_5Ge_3(X)$, $Cr_5Ge_3(X)$ $Mo_5Ge_3(X)$, and also several borides where boron is in the interstitial site, V_5Ge_3B, Nb_5Ga_3B, Nb_5Ge_3B, $Ta_5Al_3B_x$, $Ta_5Ga_3B_x$, and $Ta_5Ge_3B_x$.

References for Table XV

a. H. Nowotny, B. Lux, and H. Kudielka, *Monatsh. Chem.* **87**, 447 (1956).
b. E. Parthé, *Acta Cryst.* **12**, 559 (1959).

c. E. Parthé, H. Nowotny, and H. Schmid, *Monatsh. Chem.* **86**, 385 (1955).
d. W. Jeitschko, H. Nowotny, and F. Benesovsky, *Monatsh. Chem.* **94**, 844 (1963).
e. L. Brewer and O. Krikorian, *J. Electrochem. Soc.* **103**, 38 (1956).
f. E. Parthé, H. Schachner, and H. Nowotny, *Monatsh. Chem.* **86**, 182 (1955).
g. E. Parthé, W. Jeitschko, and V. Sadagopan, *Acta Cryst.* **19**, 1031 (1965).
h. H. Nowotny, in "Electronic Structure and Alloy Chemistry of the Transition Elements" (P. A. Beck, ed.), p. 179. New York, 1963.
i H. Nowotny, W. Jeitschko, and F. Benesovsky, *Planseeber Pulvermet.* **12**, 31 (1964).
j. W. Rieger, H. Nowotny, and F. Benesovsky, *Monatsh. Chem.* **96**, 98 (1965).

References

1. B. Aronsson, T. Lunström, and S. Rundqvist, "Borides, Silicides, and Phosphides." Wiley, New York, 1965.
2. W. B. Pearson, "A Handbook of Lattice Spacings and Structures of Metals and Alloys," Vol. I. Pergamon Press, Oxford, 1958.
2a. W. B. Pearson, "A Handbook of Lattice Spacings and Structures of Metals and Alloys," Vol. II. Pergamon Press, Oxford, 1967.
3. G. Hägg, *Z. Phys. Chem., Abt.* **B 12**, 33 (1931).
4. E. Parthé, *Z. Kristallogr., Kristallgeometric, Kristallphys., Kristallchem.* **115**, 52 (1961).
5. L. S. Darken and R. W. Gurry, "Physical Chemistry of Metals." McGraw-Hill, New York, 1953.
6. K. W. Andrews and H. Hughes, *J. Iron Steel Inst., London* **193**, 304 (1959).
7. E. Parthé and V. Sadagopan, *Monatsh. Chem.* **93**, 263 (1962).
8. J. Leciejewicz, *Acta Cryst.* **14**, 200 (1961).
9. G. Brauer and R. Esselborn, *Z. Anorg. Allg. Chem.* **309**, 151 (1961).
10. H. J. Beattie, Jr., in "Intermetallic Compounds" (J. H. Westbrook, ed.), p. 144. Wiley, New York, 1967.
11. F. Jellinek, *Oesterr. Chem.-Ztg.* **60**, 311 (1959).
12. K. Yvon, H. Nowotny, and R. Kieffer, *Monatsh. Chem.* **98**, 34 (1967).
13. E. Rudy, S. Windisch, A. J. Stosick, and J. R. Hoffman, *Trans. AIME* **239**, 1247 (1967).
14. N. S. Gorbunov, *Izv. Akad. Nauk SSSR, Otd. Khim. Nauk* **11**, 2093 (1961).
15. H. Goretzki, *Phys. Status Solidi* **20**, K141 (1967).
16. C. Froidevaux and D. Rossier, *J. Phys. Chem. Solids* **28**, 1197 (1967).
17. C. H. deNovion, R. Lorenzelli, and P. Costa, *C. R. Acad. Sci.* **263**, 775 (1966).
18. P. Costa, in "Anisotropy in Single-Crystal Refractory Compounds" (F. W. Vahldiek and S. A. Mersol, eds.), Vol. I, p. 151. Plenum Press, New York, 1968.
19. D. Rossier, Ph.D Thesis, Orsay, 1966.
20. J. D. Venables, D. Kahn, and R. G. Lye, *Phil. Mag.* [8] **18**, 177 (1968).
21. B. Holmberg, *Acta Chem. Scand.* **16**, 1255 (1962).
22. N. Schönberg, *Acta Chem. Scand.* **8**, 199 (1954).
23. G. Brauer and K. H. Zapp, *Z. Anorg. Allg. Chem.* **277**, 129 (1954).
24. For a review article, see V. I. Khitrova and Z. G. Pinsker, *Sov. Phys.—Crystallogr.* **6**, 712 (1962).
25. D. Meinhardt and O. Krisement, *Z. Naturforsch.* A **15**, 880 (1960).
26. A. Westgren, *Sv. Kem. Tidskr.* **45**, 141 (1933).

27. H. Nowotny and F. Benesovsky, in "Phase Stability in Metals and Alloys" (P. S. Rudman, J. Stringer, and R. I. Jaffee, eds.), p. 319. McGraw-Hill, New York, 1967.
28. H. Nowotny, W. Jeitschko, and F. Benesovsky, *Planseeber. Pulvermet.* **12**, 31 (1964). References 27 and 28 are excellent reviews containing many pertinent references.
29. W. Jeitschko, H. Nowotny, and F. Benesovsky, *Monatsh. Chem.* **95**, 436 and 1040 (1964).
30. L. J. Hütter and H. H. Stadelmaier, *Acta Met.* **6**, 367 (1958).
31. H. H. Stadelmaier, *Z. Metallk.* **52**, 758 (1961); this reference includes a summary of previous research on Perovskite structures by Stadelmaier and co-workers.
32. P. Rautala and J. T. Norton, *Trans. AIME* **194**, 1045 (1952).
33. N. Schönberg, *Acta Met.* **2**, 837 (1954).
34. K. Kuo, in discussion of paper by Rautala and Norton, *J. Metals* **5**, 744 (1953).
35. K. Whitehead and L. D. Brownlee, *Planseeber. Pulvermet.* **4**, 62 (1956).
36. E. Rudy, F. Benesovsky, and L. Toth, *Z. Metallk.* **54**, 345 (1963).
37. L. Toth, H. Nowotny, F. Benesovsky, and E. Rudy, *Monatsh. Chem.* **92**, 794 and 956 (1961).
38. W. Jeitschko, H. Nowotny, and F. Benesovsky, *Monatsh. Chem.* **94**, 565 (1963).
39. W. Jeitschko, *Monatsh. Chem.* **97**, 1472 (1966).
40. A. Westgren, *Jernkontorets Ann.* **117**, 1 (1933).
41. V. Adelsköld, A. Sundelin, and A. Westgren, *Z. Anorg. Allg. Chem.* **212**, 401 (1933).
42. E. N. Kislyakova, *Zh. Fiz. Khim.* **17**, 108 (1943).
43. R. Kiessling, *3rd Proc. Int. Symp. Reactiv. Solids* (1952), Vol. 2, p. 1065 (1954).
44. S. Takeda, *Sci. Rep. Tohoku Univ., Honda Anniversary Volume*, p. 864 (1936).
45. K. Kuo, *Acta Met.* **1**, 301 and 611 (1953).
46. P. Rautala and J. T. Norton, *J. Metals* **4**, 1045 (1952).
47. P. Rautala and J. T. Norton, *Plansee Proc., Pap. Plansee Semin. "De Re Metal," 1st,* 1952, p. 303 (1953).
48. R. Kiessling, discussion of a paper by P. Rautala and J. T. Norton, *J. Metals* **5**, 744 (1953).
49. A. G. Metcalfe, unpublished reports, Hard Tools Ltd., quoted in E. J. Sanford and E. M. Trent, *Iron Steel Inst., London, Spec. Rep.* **38**, 84 (1947).
50. V. I. Arkharov and S. T. Kiselev, *Izv. Akad. Nauk SSSR, Otd. Tekh. Nauk* p. 136 (1949).
51. A. Taylor and K. Sachs, *Nature* **169**, 411 (1952).
52. G. A. Yurko, J. W. Barton, and J. G. Parr, *Acta Cryst.* **12**, 909 (1959).
53. M. H. Mueller and H. W. Knott, *Trans. AIME* **227**, 674 (1963).
54. J. Leciejewicz, *J. Less-Common Metals* **7**, 318 (1964).
55. E. Parthé, W. Jeitschko, and V. Sadagopan, *Acta Cryst.* **19**, 1031 (1965).
56. H. Nowotny, in "Electronic Structure and Alloy Chemistry of the Transition Elements" (P. A. Beck, ed.), p. 179. Wiley, New York, 1963.
57. H. Nowotny, E. Parthé, R. Kieffer, and F. Benesovsky, *Monatsh. Chem.* **85**, 255 (1954).

3
Phase Relationships

I. Introduction

Although phase relationships for transition-metal carbides and nitrides have been extensively investigated for years, many of the binary phase diagrams are still inexactly known, and there exists a great deal of controversy about proposed diagrams. Most of the research has centered on the carbide phase diagrams (*1, 2*). Because of the high temperatures involved and the complexity of many of the phase relationships, a combination of sophisticated techniques and equipment including DTA, high temperature neutron diffraction, and mass spectroscopy have been brought to bear on the problem in addition to the more standard techniques of X-ray diffraction and metallography. Despite this sophistication and the large research effort considerable controversy still remains. In the past many individual investigators were studying phase relationships, but now it is clear that future research in this area can be effectively accomplished only by a few large, well-financed groups with the necessary sophisticated research tools.

The major problem in establishing phase relations is to develop techniques to cope with the very high temperatures involved. At these temperatures. interstitial phases are extremely reactive with impurities such as oxygen. As a result, many of the phase diagrams classified as binary are in fact ternary or higher order. The experiment performed by Rudy and Harmon

(*3*) on the melting point of tantalum with different impurity contents illustrates this experimental problem. The results are given in Table I.

TABLE I

MELTING POINTS OF TANTALUM AS A
FUNCTION OF IMPURITY CONTENT[a]

Total impurity content (ppm)	Melting point ($^{\circ}$C)[b]
<150	3014±10[c]
>300	3006±14
500	2985±10
900	2960±15

[a] See Rudy and Harman (*3*).
[b] Refers to the onset of melting; a two-phase alloy system melts over a range of temperatures.
[c] Error limits refer to one standard deviation and do not include systematic errors which are probably larger.

Impurities can also stabilize phases which would otherwise not exist in the binary or higher-order phase diagram. Kuo and Hägg (*4*), for example, reported the existence of γ'-MoC with a hexagonal cell and γ-MoC with the WC type structure. These phases have not been found in subsequent investigations of the MoC phase diagram by Nowotny and co-workers (*5*) or by Rudy and co-workers (*2*). Apparently γ'-MoC and γ-MoC are stabilized by small quantities of oxygen and more properly belong to the Mo–C–O ternary system (*5*). As another example, Schönberg (*6*) suggested that the $B1$ phase found by Brauer and Jander (*7*) in the Nb–N system really belonged to the Nb–N–O system. Subsequent investigations (*8*), however, showed that although the $B1$ phase does exist in the Nb–N system, it is a high-temperature phase not necessarily observed in slowly cooled specimens.

Even when care is used to reduce impurity levels to a minimum, methods of preparation can affect phase relationships. Thin films of the refractory materials provide an extreme illustration. The $B1$ phase is not stable in bulk samples of the Ta–N system prepared under 1 atm of N_2 pressure, but Schwartz (*9*) has observed this structure in reactively sputtered thin films. The impurity levels of these films are probably low, since the phase is superconducting, and the presence of oxygen in any thin film of a refractory nitride with the $B1$ structure rapidly destroys superconductivity.

In this chapter, we attempt to present up-to-date phase diagrams for carbides and nitrides and, where controversy exists, to present more than one diagram. In 1967, Storms presented a thorough evaluation of the phase diagrams and lattice parameters for the transition-metal carbides in *The Refractory Carbides* (*1*). For the most part this evaluation is still valid. For Ti–C, Zr–C, Hf–C, and Ta–C Storms' evaluation is still preferred because little additional information has appeared which would tend to outdate it. For several binary systems, however, considerable additional information has appeared. In many cases the nature of this new research is so controversial that a new evaluation did not seem possible without still further research to resolve the conflicting points. In these situations, both sides of the conflicting views are presented. For the nitrides, the phase information is not as well defined as for the carbides and an attempt was made to bring together some of the existing diagrams.

In addition to the more recent reviews by Storms (*1*) and Rudy (*2*), other reviews of the phase relationships of carbides and nitrides have been compiled by Hansen *et al.* (*10*), Schwarzkopf and Kieffer (*11*), Kieffer and Benesovsky (*12*), Storms (*13*), and Samsonov and Umanskiy (*14*).

II. Phase Diagrams of Carbides

IIA. Fourth Group

The phase diagrams of the fourth-group transition metals with carbon are very similar. They are characterized by only one carbide phase, a monocarbide with the $B1$ structure with a wide homogeneity range. All $B1$ phases of the fourth-group carbides melt congruently at slightly substoichiometric compositions and all form a eutectic with carbon at higher carbon concentrations.

Each of the fourth-group metals, Ti, Zr, and Hf, undergoes an allotropic transformation in which the low-temperature α phase with the hcp($A3$) structure transforms to the high-temperature β-bcc($A2$) structure. According to Storms (*1*), the allotropic transformation temperatures for Ti, Zr, and Hf are, respectively, 882.2 ± 0.5, 865, and 1740°C. The solubility of carbon is limited in Ti and Zr, but is extensive in Hf. The low-temperature phase α-Hf is stabilized to much higher temperatures by the interstitial solution. At 2360°C, α-Hf with 14 at. % of carbon in solution decomposes peritectically into HfC_{1-x} and liquid.

Phase diagrams for the fourth-group carbides have been reviewed by Storms and his diagrams for Ti–C and Zr–C are given in Figs. 1 and 2. Rudy's diagram for Hf–C is given in Fig. 3. In general, the phase diagrams

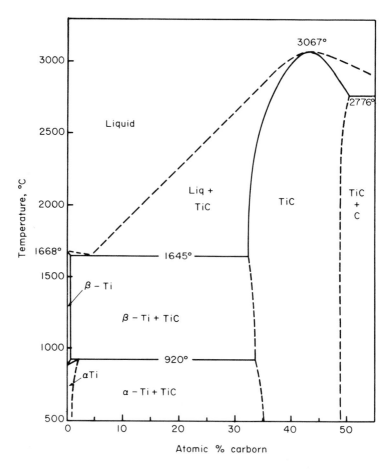

Fig. 1. Proposed phase diagram for titanium–carbon. [After Storms' evaluation (1).]

established by different investigators are in fair agreement (1, 2, 15–18). For example, the diagram for Zr–C determined by Sara *et al.* (15, 16) is almost identical to that of Rudy *et al.* (17).

IIB. Fifth Group

The phase diagrams for the fifth-group carbides V–C, Nb–C, and Ta–C are more complex than those for the fourth group. They are characterized by the presence of an Me$_2$C phase in addition to the monocarbide. The

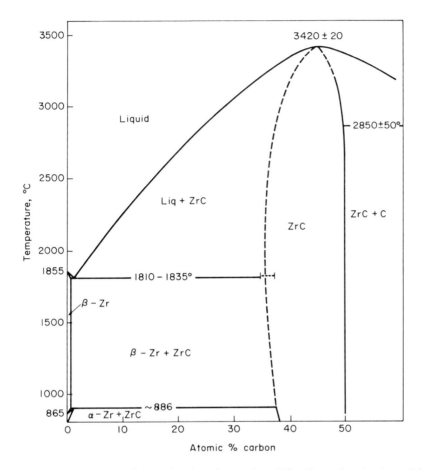

Fig. 2. Proposed phase diagram for zirconium–carbon.[After Storms' evaluation (1).]

Nb–C and Ta–C systems are probably similar. They differ from the V–C system primarily in composition range and melting point of the mono-carbide. There are at least two Me_2C phases in the V–C, Nb–C, and Ta–C systems. These phases differ in the ordering of carbon atoms (see Chapter 2). The high-temperature β-V_2C, γ-Nb_2C, and β-Ta_2C phases are probably the disordered L_3' structures. In each system, the Me_2C phase decomposes peritectically to form liquid and monocarbide, and there is a eutectic re-action between the low-carbon-content liquid and the Me and Me_2C phases. There is a second eutectic reaction between the monocarbide and graphite phases. Latest studies on the V–C system show that VC_{1-x} melts

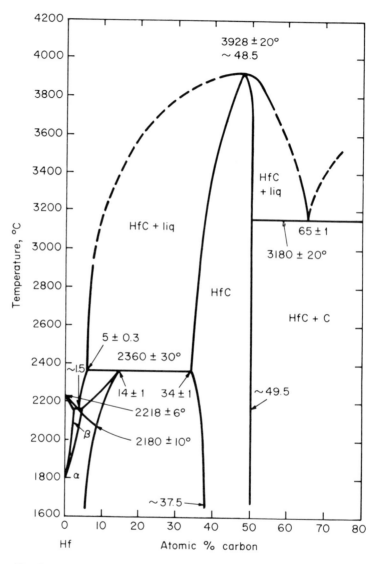

Fig. 3. Proposed phase diagram for hafnium–carbon. [After Rudy (2).]

congruently rather than decomposes peritectically as was previously supposed (19–21). Besides the two intermediate phases, MeC and Me_2C, there
is a metastable phase in each phase diagram at the approximate composition
Me_3C_2, the so-called Brauer ζ phase (22). The crystal structure of this phase

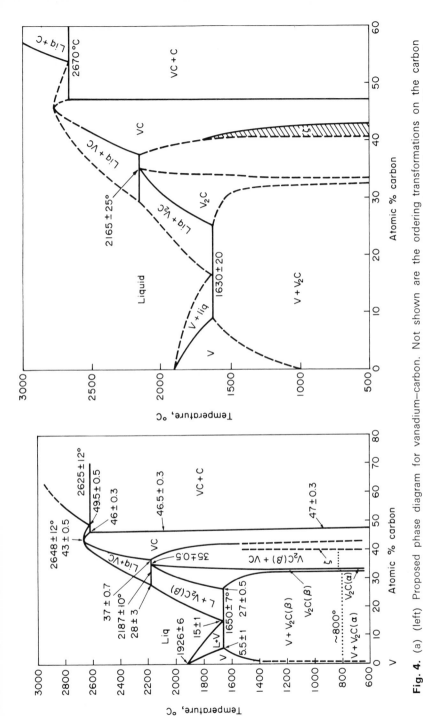

Fig. 4. (a) (left) Proposed phase diagram for vanadium–carbon. Not shown are the ordering transformations on the carbon sublattice in VC$_{1-x}$. [After Rudy et al. (2, 19).] (b) (right) Proposed phase diagram for vanadium–carbon. [After Storms (1) as modified by Adelsberg and Cadoff (21).]

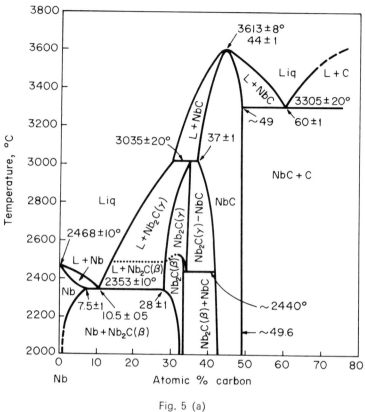

Fig. 5 (a)

is unknown. It probably forms as the result of a nonequilibrium epitaxial precipitation from the carbon-defective monocarbide (23).

The proposed diagrams for V–C shown in Fig. 4 do not include the ordering of carbon atoms in the monocarbide. Recent nuclear magnetic resonance studies, discussed in Chapter 2, indicate that the carbon atoms in the VC_{1-x} phase are ordered. At V_8C_7 a cubic symmetry exists; at V_6C_5, a hexagonal symmetry. Furthermore, studies on temperature dependence of the mechanical properties of VC, as discussed in Chapter 5, indicate that ordering disappears above 1200°C. Neutron diffraction studies on the VC_{1-x} phase might confirm this suspected ordering and correct the phase diagram if more than one phase is found in this composition region. Impurity levels would have to be carefully controlled. Since substoichiometric VC_{1-x} is very reactive with oxygen, this ordering, or lack of it if the $B1$ phase does exist, may result from the solution of oxygen.

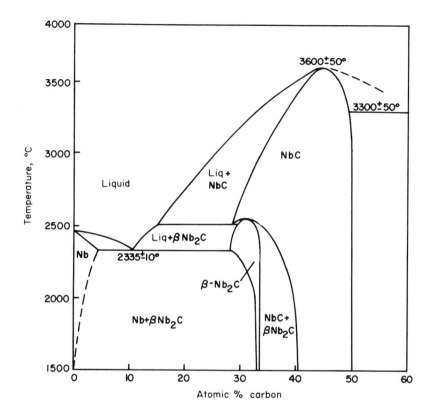

Fig. 5. (a) (opposite) Proposed phase diagram for niobium–carbon. Not shown are an additional order transformation in Nb_2C at 1230°C and the Brauer ζ phase at the approximate composition Nb_3C_2. [After Rudy *et al.* (*2, 19*).] (b) (above) Proposed phase diagram for niobium–carbon. [After Storms *et al.* (*24*).]

It is clear from the present V–C diagram that VC_{1-x} is never stable beyond the composition $VC_{0.88}$. This behavior contrasts with that of monocarbides NbC_{1-x} and TaC_{1-x} and the fourth-group monocarbides. The sixth-group carbides MoC and WC with the *B*1 structure also form only with large deviations from the stoichiometric one-to-one ratio. If the stoichiometric one-to-one compound does form, the metal atoms will no longer form in a close-packed structure.

There is considerable controversy about the phase relationships in Nb–C (Figs. 5a and 5b). Storms *et al.* (*24*) have recently measured the vapor pressure of Nb in the monocarbide as a function of composition and temperature using Knudsen effusion in a mass spectrometer and they find that the vapor pressure measurements are in conflict with existing phase diagrams (such as Fig. 5a). They therefore proposed a new diagram (Fig. 5b) in which Nb_2C decomposes at a much lower temperature into liquid and NbC and in which NbC_{1-x} exists at high temperatures with more carbon deficiency than previously supposed. In contrast, Rudy's diagram (Fig. 5a) shows that instead of Nb_2C peritectically decomposing at ~2500°C it undergoes another crystal structure change, probably transforming from β-Nb_2C (ϵ-Fe_2N type) to γ-Nb_2C (L_3' type). Rudy, however, was never able to quench in the γ-Nb_2C phase and could only detect its presence or some other phases' presence by DTA measurements. These differing opinions could be effectively resolved using high temperature X-ray or neutron diffraction techniques. Because of the similarities of the phase relationships between Nb–C and Ta–C (see Fig. 6) and also to a lesser extent to V–C, some additional doubt is placed on the correctness of the proposed diagrams for the latter systems.

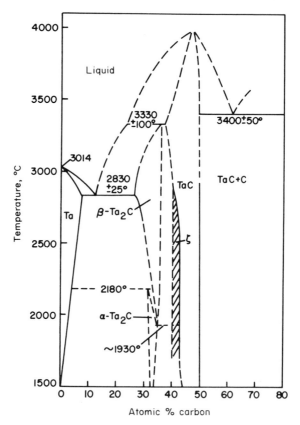

Fig. 6. Proposed phase diagram for tantalum–carbon. [After Storms' evaluation (1).]

Table II lists crystal structures and phase relationships in the fifth-group phase diagrams. The phase designations and transformation temperatures are those of Rudy *et al.* (*19*).

IIC. Sixth Group

The phase diagrams of the sixth-group carbides, Cr–C, Mo–C, and W–C, are more varied and somewhat more complex than those of the fourth or fifth group. In the Cr–C system, the metal atoms no longer form close-packed structures. The Cr–C system is entirely unrelated to that of Mo–C or W–C or to the other diagrams for the fourth and fifth-group carbides. This phase diagram is shown in Fig. 7 (*2*, *25*). Table III lists the crystal structures of the intermediate phases, $Cr_{23}C_6$, Cr_7C_3, and Cr_3C_2.

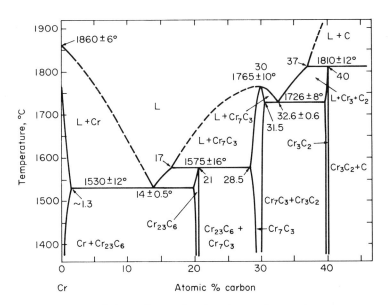

Fig. 7. Proposed phase diagram for chromium–carbon. [After Rudy (*2*).]

There have been numerous experimental investigations and evaluations of the Mo–C and W–C systems (*1*, *2*, *26–29*). Both diagrams are extremely difficult to establish because several of the phases are stable only at very high temperatures. These phases can be retained at room temperature only by drastic quenching techniques. As a result, some of the high-temperature composition boundaries are rather uncertain.

TABLE II

STRUCTURES AND LATTICE PARAMETERS OF FIFTH-GROUP CARBIDES

Phase	Designation (a)	Structure type	Lattice parameter (Å)	Ref.
V_2C	α-V_2C ($T < 800°C$)	Orthorh ζ-$Fe_2N^{a,\,b}$	$a = 4.577; b = 5.742;$ $c = 5.037$	(b)
			$a = 11.49; b = 10.06;$ $c = 4.55$	(c)
	β-V_2C ($T > 800°C$)	Hexag L_3'	$a = 2.885; c = 4.570$ at $VC_{0.47}$	(d)
			$a = 2.902; c = 4.577$ at $VC_{0.50}$	(d)
ζ-VC_{1-x}	(At ~40 at. % C)	Unknown		
VC_{1-x}	—	Cubic $B1^c$	$a = 4.131$ at $VC_{0.73}$	(d)
			$a = 4.166$ at $VC_{0.87}$	(d)
Nb_2C	α-Nb_2C ($T < 1230°C$)	Orthorh ζ-Fe_2N^a	$a = 10.92; b = 4.974;$ $c = 3.090$	(e)
			$a = 12.36; b = 10.855;$ $c = 4.968$	(c)
	β-Nb_2C ($1230 < T < 2500°C$)	Hexag probably ϵ-Fe_2N	$a = 5.407; c = 4.974$	(f)
	γ-Nb_2C ($T > 2500°C$)	Hexag L_3'	$a = 3.127; c = 4.965$ at $NbC_{0.49}$	(g)
			$a = 3.127; c = 4.972$ at $NbC_{0.50}$	(g)
ζ-NbC_{1-x}	(At ~40 at. % C)	Unknown		(h)
NbC_{1-x}	—	Cubic $B1$	$a = 4.431$ at $NbC_{0.71}$	(g)
			$a = 4.470$ at $NbC_{0.99}$	(g)

Ta$_2$C	α-Ta$_2$C ($T < 2180°C$)	Hexag C6^d	$a = 3.100$; $c = 4.931$ at TaC$_{0.46}$ $a = 3.102$; $c = 4.940$ at TaC$_{0.50}$	(i) (i)
	β-Ta$_2$C ($T > 2180°C$)	Hexag L$_3'$	$a \cong 3.102$; $c \cong 4.940$	(i)
ζ-TaC$_{1-x}$	(At. ~40 at. % C)			(j)
TaC$_{1-x}$	—	Cubic B1	$a = 4.412$ at TaC$_{0.74}$ $a = 4.456$ at TaC$_{0.99}$	(i) (i)

[a] There is some discrepancy about the crystal structures of α-V$_2$C and α-Nb$_2$C. The structure proposed by Rudy and Brukl (Ref. c) for both phases is similar to that of ζ-Fe$_2$N (Ref. b) but with b and c axes doubled.

[b] There is also an ε-V$_2$C reported with the ε-Fe$_2$N structure (see Chapter 2, Fig. 6).

[c] The carbon atoms in this structure are known to order at V$_8$C$_7$ and V$_6$C$_5$ (see Chapter 2).

[d] Lattice parameters given at L$_3'$.

References for Table II

a. E. Rudy, S. Windisch, and C. E. Brukl, *Planseeber. Pulvermet.* **16**, 3 (1968).

b. K. Yvon, W. Rieger, and H. Nowotny, *Monatsh. Chem.* **97**, 689 (1966).

c. E. Rudy and C. E. Brukl, *J. Am. Ceram. Soc.* **50**, 265 (1967).

d. E. K. Storms and R. J. McNeal, *J. Phys. Chem.* **66**, 1401 (1962).

e. K. Yvon, H. Nowotny, and R. Kieffer, *Monatsh. Chem.* **98**, 34 (1967).

f. N. Terao, *Jap. J. Appl. Phys.* **3**, 104 (1964).

g. E. K. Storms and N. H. Krikorian, *J. Phys. Chem.* **64**, 1471 (1960).

h. G. Brauer and R. Lesser, *Z. Metallk.* **50**, 8 (1959).

i. E. Rudy and D. P. Harmon, AFML-TR-65-2, Part I, Vol. V (1966).

j. R. Lesser and G. Brauer, *Z. Metallk.* **49**, 622 (1958).

TABLE III

PHASE RELATIONSHIPS IN THE Cr–C SYSTEM

Phase	Crystal structure	Lattice parameter (Å)
$Cr_{23}C_6$	Complex fcc with 116 atoms per unit cell ($D8_4$ type)	$a = 10.655$ (a)
Cr_7C_3	Hexagonal with 80 atoms per unit cell	$a = 14.01$ (a) $c = 4.525$
Cr_3C_2	Orthorhombic with 20 atoms per unit cell (D_{2h}^{16}-Pbnm) (b)	$a = 11.47$ (a) $b = 5.545$ $c = 2.830$

References for Table III

a. P. Stecher, F. Benesovsky, and H. Nowotny, *Planseeber Pulvermet.* **12**, 89 (1964).
b. Carbon positions have been determined by neutron diffraction. D. Meinhardt and O. Krisement, *Z. Naturforsch.* A **15**, 880 (1960).

In the Mo–C system, there are at least four well-established intermediate phases α-Mo_2C, β-Mo_2C, η-MoC_{1-x}, and α-MoC_{1-x} (2). The high-temperature β-Mo_2C probably has the disordered L_3' structure, whereas the α-Mo_2C phase has an orthorhombic structure in which the carbon atoms are ordered. According to Rudy *et al.* (26) the transformation between α-Mo_2C and β-Mo_2C is first order only at the composition 32.5 at. % C. At compositions poorer in carbon, the transformation occurs over a temperature range by a displacive reaction (diffusion aided). Bowman *et al.* (30), who studied the transformation with the aid of high temperature neutron diffraction, find that the reactions

$$Mo_2C(\beta) \rightleftarrows Mo_2C(\alpha) + Mo$$

and

$$Mo_2C(\beta) \rightleftarrows Mo_2C(\alpha) + C$$

appear to be eutectoidal and isothermal (first order). Their results, therefore, do not support those of Rudy *et al.* (26).

Although possible ordering of the carbon atoms in the η-MoC_{1-x} phase has not been explored with neutron diffraction techniques, such ordering is likely to occur, since the composition of this phase is nearly Mo_3C_2 and not all the octahedral carbon sites are equivalent. In fact, of the available sites for the carbon atoms, two thirds are equivalent positions and one third is slightly different.

In addition to these four well-established phases, Kuo and Hägg (*4*) have reported the existence of γ-MoC with the WC-type structure and γ′-MoC with the TiP-type structure. Both phases are probably stabilized by oxygen or other impurities (*5*).

In the W–C system, there are four intermediate phases: α-W$_2$C, β-W$_2$C, β-WC$_{1-x}$, and α-WC (*27*). The two W$_2$C phases differ in arrangement of the carbon atoms; β-W$_2$C cannot be retained in its pure state at room temperature. Its most probable structure is L_3'. There is, however, some difference of opinion about the existence of β-W$_2$C (*29, 30*). β-WC$_{1-x}$ has the $B1$ structure, and WC has the WC hexagonal structure, in which W atoms no longer form a close-packed structure. The phase diagrams for the Mo–C and W–C systems are shown in Figs. 8 and 9. Table IV lists the crystal structures of the intermediate phases.

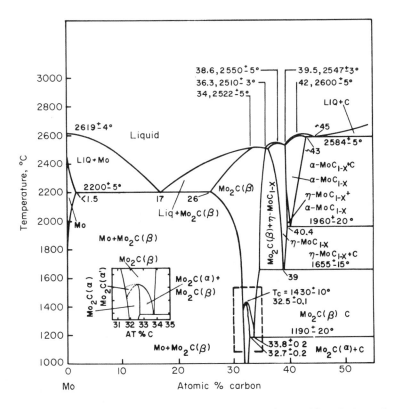

Fig. 8. Proposed phase diagram for molybdenum–carbon. [After Rudy and co-workers (*2, 26*).]

TABLE IV. CRYSTAL STRUCTURES AND LATTICE PARAMETERS IN THE Mo–C AND W–C SYSTEMS

System	Phase	Crystal structure	Crystal structure illustrated at this fig. number in Chapter 2	Lattice parameters (Å)
Mo–C	α-Mo$_2$C	ζ-Fe$_2$N type, orthorhombic (a)	5, 6	$a = 7.244$, $b = 6.004$, $c = 5.199$ (a)
	β-Mo$_2$C	Probably L_3' with random occupancy of the carbon sites, hexagonal (b)	2, 5, 6	$a = 2.998$–3.012 (b, c), $c = 4.733$–4.778 (b, c)
	η-MoC$_{1-x}$	η-MoC type, hexagonal (d)	2	$a = 3.006$, $c = 14.61$ (d)
	α-MoC$_{1-x}$	B1, cubic (e, f)	2	$a = 4.27$ (e), $a = 4.281$ (f)
	γ-MoC	WC type (g)	2, 3b	$a = 2.898$, $c = 2.809$ (g)
	γ'-MoC	TiP type (g)	2	$a = 2.932$, $c = 10.97$ (g)
W–C	α-W$_2$C	C6 type, hexagonal (h)	5, 6	$a = 2.985$, $c = 4.717$ at 29.2 at. % C (i); $a = 3.001$, $c = 4.728$ at 33.3 at. % C (i)
	β-W$_2$C	Probably L_3' with random occupancy of the carbon sites, hexagonal (i)	2, 5, 6	$a = 3.002$, $c = 4.75$–4.76 (i)
	α-WC$_{1-x}$	B1, cubic (i)	2	$a = 4.220$ at $\sim$38 at. % C (i)
	WC	WC type (j)	2, 3b	$a = 2.906$, $c = 2.837$ (i)

References for Table IV

a. E. Parthé and V. Sadagopan, *Acta Cryst.* **16**, 202 (1963).
b. E. Rudy, S. Windisch, A. J. Stosick, and J. R. Hoffman, AFML-TR-65-2, Part I, Vol. XI (1967).
c. H. Nowotny and R. Kieffer, *Z. Metallk.* **38**, 257 (1947).
d. H. Nowotny, E. Parthé, R. Kieffer, and F. Benesovsky, *Monatsh. Chem.* **85**, 255 (1954).
e. E. V. Clougherty, K. H. Lothrop, and J. A. Kafalas, *Nature* **191**, 1194 (1961).
f. E. Rudy and F. Benesovsky, *Planseeber. Pulvermet.* **10**, 42 (1962).
g. K. Kuo and G. Hägg, *Nature* **170**, 245 (1952).
h. L. N. Butorina and Z. G. Pinsker, *Sov. Phys.—Crystallogr.* **5**, 560 (1961).
i. E. Rudy, S. Windisch and J. R. Hoffman, AFML-TR-65-2, Part I, Vol. VI (1966).
j. E. Parthé and V. Sadagopan, *Monatsh. Chem.* **93**, 263 (1962).

III. Phase Diagrams of the Nitrides

Information on phase diagrams of transition-metal nitrides is very sparse, frequently unreliable, and often contradictory. One particular problem confounding phase-diagram studies is the sensitivity of phase stabilities, phase boundaries, and lattice parameters to methods of preparation. It has been

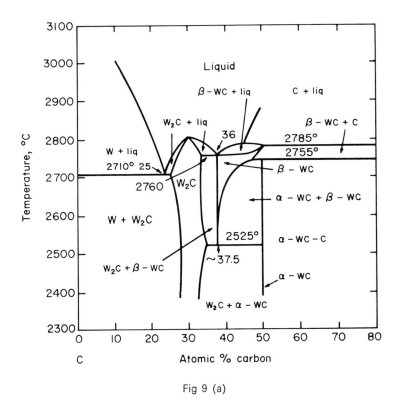

Fig 9 (a)

found, for example, that nitriding fourth-group halogens at low temperatures in ammonia produces a $B1$ structured mononitride Ti_xN and Zr_xN where $x < 1$ (*31, 32*). Density studies show that in these phases the metal lattice is predominantly defective. Nitriding $Ti(TiH_2)$ and $Zr(ZrH_2)$ in nitrogen or ammonia, however, produces a $B1$ structured mononitride TiN_x or ZrN_x in which $x < 1$; that is, the nitrogen sublattice is predominantly defective.

Thin nitride films also frequently appear to have different crystal structures from those of the bulk material. TaN formed as a thin film by reactive

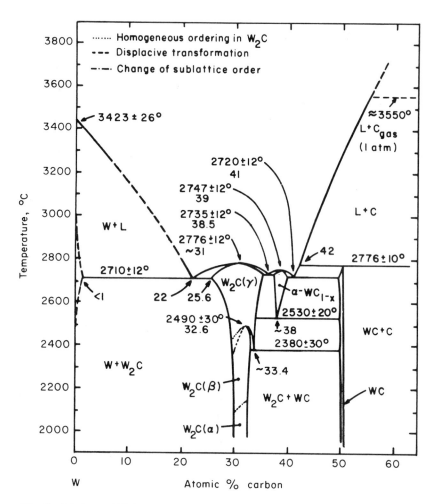

Fig. 9. (a) (page 85) Phase diagram for tungsten–carbon. Here β-WC refers to the *B*1 phase. [After Sara (*29*).] (b) (above) Proposed phase diagram for tungsten–carbon. [After Rudy (*2*).]

sputtering has the *B*1 structure (*9*), whereas bulk TaN does not have this structure. Thin films of NbN (*33, 34*) and Nb_2N (*35*) have been prepared, but several additional phases found in bulk samples have not been observed in these thin films.

Other problems in determining phase diagrams of nitrides are the high evaporation rates of nitrogen and the difficulties of obtaining accurate chemical analyses of nitrogen and impurity levels. While stated accuracies on nitrogen determinations may be within 1%, the best usually obtained is actually 2–3%, and still higher errors are frequently encountered. The effect of impurities such as oxygen is difficult to determine. Oxygen forms solid solutions with the nitrides and also forms the oxides. Therefore, determination of the amount of oxygen present in the solid solution is difficult.

IIIA. Fourth-Group Nitrides

The phase diagram for Ti–N is incompletely known. Two intermediary phases, TiN($B1$) and ϵ-Ti$_2$N, are known to exist (32, 36–42). TiN has a broad composition range from about TiN$_{0.6}$ to about TiN$_{1.0}$ at 1400°C. If TiN is prepared at low temperatures by reacting NH$_3$ with TiCl$_4$, a composition range of TiN to TiN$_{1.16}$ results (32).

Figure 10 shows the variation of lattice parameter with composition for TiN. The lattice parameter is maximum at the stoichiometric composition.

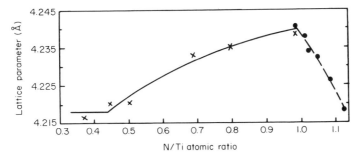

Fig. 10. The lattice parameter of the $B1$ phase TiN$_{1-x}$ is a maximum at the stoichiometric composition and decreases rapidly in overly stoichiometric compositions. [After Ehrlich (37) and Brager (32).]

Density studies in conjunction with lattice parameter studies show that at substoichiometric compositions, the nitrogen sublattice is predominantly defective, and at hyperstoichiometric compositions, the Ti sublattice is defective. Density studies also show, however, that even at the stoichiometric composition, TiN has a large fraction of vacant sites on both the metal and the nonmetal sublattices (37). Theories to predict this vacancy concentration from known thermodynamic values are discussed in the next chapter, and a possible explanation of this effect is presented in Chapter 8 on bonding and band structure.

The phase ϵ-Ti$_2$N has a tetragonal structure; $a = 4.9452$ Å and $c = 3.0342$ Å (39). This crystal structure is discussed in Chapter 2, Fig. 10.

Palty *et al.* (*36*) investigated most of the Ti–N phase relationships. The composition limits for the phase stability which these authors reported are in poor agreement with subsequent measurements. Figure 11 attempts to

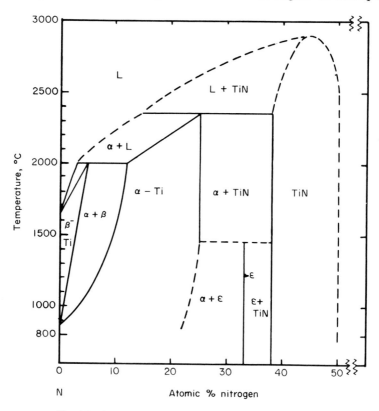

Fig. 11. Proposed phase diagram for titanium–nitrogen.

piece together current literature on these phase relationships. The only certain features are the existence of TiN and ε-Ti$_2$N and the large terminal solubilities of nitrogen into α- and β-Ti. Exact transition temperatures or composition limits are very uncertain. Not shown in the figure is the reported (*38*) ordering of nitrogen atoms in α-Ti at about 20 at. % N which results in the anti-CdI$_2$(*C*6) type structure (see Fig. 6 of Chapter 2). This reported ordering has not, however, been confirmed.

The phase diagram for Zr–N is not well established. Storms (*43*) has reviewed the earlier literature on Zr–N and summarized the information into a probable phase diagram (Fig. 12). Only one intermediary phase, ZrN

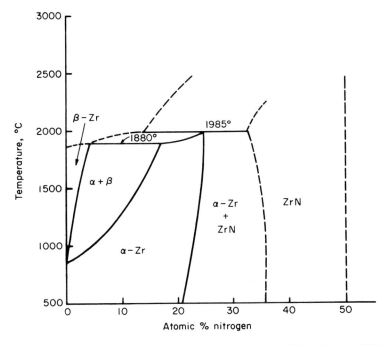

Fig. 12. Proposed phase diagram for zirconium–nitrogen. [After Storms (*43*).]

with the *B*1 structure, is known to exist. Its composition range is broad, and its stability limits depend on the manner of preparation. Prepared by nitriding Zr(ZrH$_2$) in N$_2$ or NH$_3$, ZrN$_x$ is stable only when $x \leq 1$, but when prepared by nitriding ZrI$_4$ with NH$_3$ at 750°C, Zr$_x$N is stable with $x < 1$ (*31*).

The lattice parameter of ZrN has an unusual variation with composition. At substoichiometric compositions the lattice expands slightly as nitrogen is removed from the lattice (*44–46*). No explanation has been offered for this effect.

The information on the phase relationships in the Hf–N system is very sparse. According to Rudy and Benesovsky (*46*), the only intermediary phase is HfN(*B*1), which extends from HfN$_{0.74}$ to HfN$_{1.13}$. *α*-Hf dissolves nitrogen at 1700°C to a composition limit of HfN$_{0.41}$. Overly stoichiometric HfN(*B*1) can be prepared by nitriding HfH$_2$ in N$_2$ or NH$_3$ at temperatures lower than 1500°C. (This is not the case for ZrN and TiN.) The upper limit of composition range is sensitive to both nitriding pressure and temperature. At 1 atm of N$_2$ pressure and 1000°C, an upper limit of about HfN$_{1.10}$ is reached. At 1250°C, the upper limit is about HfN$_{1.05}$ (*47*). The variation of the lattice parameter with composition (Fig. 13) (46–48) indicates that the Hf sublattice is defective when the N/Hf ratio is greater than one. The variation at substoichiometric compositions shows the same unusual behavior as does ZrN.

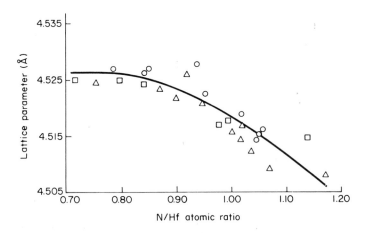

Fig. 13. Variation of the lattice parameter with composition in HfN_{1-x}. □: data from Rudy and Benesovsky (46); △: data from Lee and Toth (47); ○: data from Giorgi et al. (48).

IIIB. Fifth-Group Nitrides

The phase relationships of the fifth-group nitrides are better characterized than those of the fourth group, primarily because of the careful studies of Brauer and co-workers.

(1) Vanadium–nitrogen. Two intermediate phases exist in this system: the hexagonal β phase at a composition of about V_2N, and the $B1$ mononitride or δ phase (49, 50). The nitrogen-saturated limit of the δ phase is $VN_{1.00}$, and the vanadium-saturated limit is between $VN_{0.74}$ and $VN_{0.71}$ (49–51). The homogeneity limits for the β phase are more uncertain. The nitrogen-saturated composition limit of β is $VN_{0.49}$ according to Brauer and Schnell (49), but $VN_{0.43}$ according Hahn (50). The vanadium-saturated limit of β is between $VN_{0.35}$ and $VN_{0.40}$ (49, 50).

β-V_2N has the hexagonal ϵ-Fe_2N-type structure (52) (see Fig. 6 of Chapter 2) with $a = 4.910$ and $c = 4.541$ Å (50) at the nitrogen-rich boundary. This phase has also been referred to as ϵ-V_3N_{1+}. The structural determination of the latter phase may be influenced by the solution of hydrogen during preparation (52).

The lattice parameter of δ-VN($B1$) varies in a nearly linear fashion from $a = 4.1398$ Å at $VN_{1.00}$ to $a = 4.0662$ Å at $VN_{0.72}$ (49).

(2) *Niobium–nitrogen.* The niobium–nitrogen phase diagram is very complex and has proven experimentally difficult to establish accurately. The crystal structure relationships in this system are also interesting in that a wide variety of interrelated types are exhibited. Figure 14 summarizes three

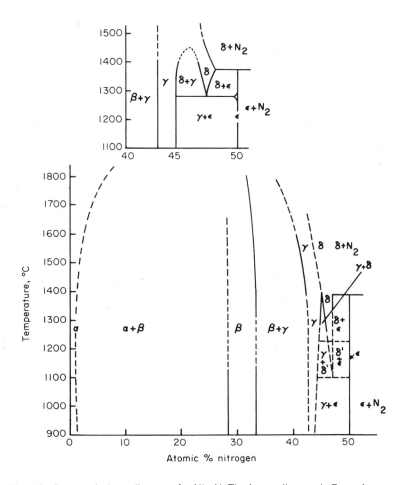

Fig. 14. Proposed phase diagrams for Nb–N. The lower diagram is Brauer's proposal given originally by Brauer and Esselborn (53) and then modified by him (54). The subsequent modification by Guard et al. (54) of the nitrogen-rich side is shown in the upper diagram.

TABLE V. CRYSTAL STRUCTURES AND LATTICE PARAMETERS IN THE Nb–N SYSTEM

Phase	Crystal structure	Crystal structure illustrated at this fig. number in Chapter 2	Lattice parameter (Å)
β-Nb$_2$N	Probably L' with random occupancy of nitrogen sites, hexagonal (a)	2, 5, 6	$a = 3.056$ at N-poor boundary (b) $c = 4.955$ $a = 3.056$ at N-rich boundary $c = 4.996$
γ-NbN$_x$	Distorted $B1$ (b) Tetragonal	—	$a = 4.385$ at N-poor boundary (c) $c = 4.310$ $a = 4.386$ at N-rich boundary $c = 4.335$
δ-NbN	$B1$ Cubic (b)	2	See Fig. 16
δ'-NbN$_x$	Anti-NiAs (b d,) Hexagonal	2, 3a	$a = 2.968$ (d) $c = 5.535$
ϵ-NbN	Metal atoms like those in TiP-type, by N stacking may be AAA. Disorder may also exist in metal atom layers (b, c, d)	2	$a = 2.958$ (e) $c = 11.272$

References for Table V

a. G. Brauer. *Z. Elektrochem.* **46**, 397 (1940).
b. G. Brauer and J. Jander, *Z. Anorg. Allg. Chem.* **270**, 160 (1952).
c. G. Brauer and R. Esselborn, *Z. Anorg. Allg. Chem.* **309**, 151 (1961).
d. N. Schönberg, *Acta Chem. Scand.* **8**, 208 (1954).
e. R. W. Guard, J. W. Savage, and D. G. Swarthout, *Trans. AIME* **239**, 643 (1967).

of the most recently proposed diagrams (*53, 54*). Table V summarizes the crystal structure types and lattice parameters.

There is some confusion about the symbol designations of these phases (*53, 55, 56*). Table VI lists the different designations previously used. In this test the notation of Brauer and Esselborn (*53*) is preferred.

The lattice parameter of the hexagonal (L_3' type or ordered derivative structure) β phase shows an unusual variation with nitrogen composition

TABLE VI

SYMBOL DESIGNATIONS FOR PHASES IN THE Nb–N SYSTEM

Brauer–Esselborn (*53*)	Brauer–Jander (*55*)	Schönberg (*56*)
α-Phase (Nb terminal solution)	Nb-Phase	α-Phase
β-Phase	Nb_2N	β-Phase
γ-Phase	Nb_4N_3	—
δ-Phase	NbN(III)	—
δ'-Phase	NbN(II)	δ-Phase
ϵ-Phase	NbN(I)	ϵ-Phase

in the range $NbN_{0.40}$ to $NbN_{0.50}$. The lattice constant a is independent of composition, but c rapidly increases between $NbN_{0.48}$ and $NbN_{0.50}$, while remaining relatively constant between $NbN_{0.40}$ and $NbN_{0.48}$ (see Fig. 15) (*53*).

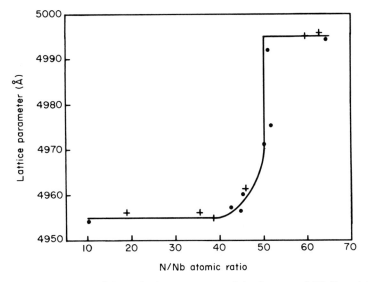

Fig. 15. Dependence of the c–lattice parameter of the hexagonal Nb_2N and N/Nb ratio. [After Brauer and Esselborn (*53*).]

γ-NbN$_x$ has a distorted NaCl structure. With increasing temperature and nitrogen concentration, the c/a ratio of the tetragonal cell approaches 1, so it is likely that γ and δ becomes one continuous phase at temperatures above 1400°C (see Figs. 14b and c).

By heating δ-NbN in high nitrogen pressures (to 240 atm), it is possible to increase the range of stability of this phase to NbN$_{1.062}$ (57). Analysis of the lattice parameter variation with composition (see Fig. 16) and density studies

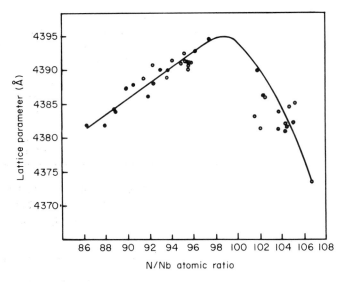

Fig. 16. Dependence of the lattice parameter of δ-NbN with N/Nb ratio. [After Brauer and Kirner (57).]

show that at the nitrogen-poor side of stoichiometry, the nitrogen sublattice is predominantly defective, while on the nitrogen-rich side of stoichiometry, the metal sublattice is defective (57). Even at stoichiometry, there is evidence to suggest a considerable number of defects on both sublattices. The $B1$ phase in the Ti–N and Hf–N systems show an analogous behavior.

There is some uncertainty about the crystal structure of ϵ-NbN. Schönberg (56) suggests that the phase has the TiP-type structure, but Brauer and Esselborn (53) suggest that the sequential ordering of N atoms may be —AAAA— (positions $0\,0\,0$, $0\,0\,\frac{1}{4}$, $0\,0\,\frac{1}{2}$, $0\,0\,\frac{3}{4}$ versus Schönberg's $0\,0\,0$, $0\,0\,\frac{1}{2}$, $\frac{1}{3}\,\frac{2}{3}\,\frac{3}{4}$, $\frac{2}{3}\,\frac{1}{3}\,\frac{1}{4}$). Brauer and Esselborn also suggest that there may be some disorder in the sequential arrangement of the Nb close-packed layers.

δ'-NbN appears to be a metastable phase which occurs only during the transition $\delta \rightarrow \epsilon$ (54).

A hexagonal NbN_x phase with the WC structure reported by Schönberg (56) has not been confirmed.

(3) Tantalum–nitrogen. Schönberg (58) and Brauer and co-workers (59) investigated the Ta–N phase relationships. Their investigations agree on the following data:

(1) Nitrogen is soluble in β-Ta to about $TaN_{0.05}$.

(2) A Ta_2N phase exists in which the metal atoms are arranged as hcp. The composition range of this phase is from $TaN_{0.41}$ to $TaN_{0.50}$ with the corresponding lattice parameters at each limit $a = 3.042kX$, $c = 4.905kX$ and $a = 3.042kX$, $c = 4.909kX$ (59).

(3) A TaN phase exists (ϵ-TaN) which has the $B35$-type structure [$a = 5.185$, $c = 2.908$ Å (58)]. This structure was discussed in Chapter 2 (see Section IIIB). The composition range of this phase is very narrow.

Less certain is the existence of superlattice lines in β-Ta and the existence of a δ-$TaN_{0.80}$—$TaN_{0.90}$ phase with the WC-type structure, both of which were reported by Schönberg (58).

Earlier reports (60) about the existence of Ta_3N_5 have been confirmed (61). The crystal structure of this phase is still unknown.

IIIC. Sixth-Group Nitrides

Little is known about the sixth-group nitrides. This lack of information stems from two causes: (1) the sixth-group nitrides are characterized by chemical instability, and (2) the compounds are not generally considered to be refractory because of their rapid dissociation to N_2 and the pure element at high temperatures. Nevertheless, the sixth-group nitrides are interesting for theoretical reasons. Their chemical instability is clearly related to a filling of antibonding electron states. The situation is somewhat analogous to that of the carbides of Mo and W, which also have relatively low melting points. These concepts will be discussed in Chapter 8 on bonding.

The sixth-group nitrides become progressively less stable in the order Cr to Mo to W. Tungsten, for example, will not form a stable nitride in nitrogen gas at any temperature and will react only very slowly with NH_3 at 600–800°C.

For all sixth-group elements, the terminal solubility of N_2 is slight.

(1) Chromium–nitrogen. Two intermediate phases are known to exist in the chromium–nitrogen system—ϵ-Cr_2N and CrN. ϵ-Cr_2N is stable from about $CrN_{0.38}$ to $CrN_{0.50}$ (62). The composition of CrN is uncertain, but it is probably variable in nitrogen content (63).

The atomic positions of the Cr atoms in ϵ-Cr_2N were determined by Blix

(64) to be hcp. Eriksson *(62)*, however, determined the positions of the N atoms as well as the Cr atoms, and suggested a superstructure with $a' = \sqrt{3}a_{\text{Blix}}$ and $c' = c$. The unit cell size is $a = 4.750$, $c = 4.429kX$ at the N-poor boundary and $a = 4.796$, $c = 4.470kX$ at the N-rich boundary.

CrN has the NaCl(*B*1) crystal structure. The reported lattice parameters vary somewhat, and no correlations between composition and lattice parameter are available. The value is about 4.149 Å *(65)*.

(2) Molybdenum–nitrogen. Hägg *(66)* investigated the Mo–N phase diagram and reported three intermediate phases: a β phase with a face-centered tetragonal structure near 28 at. % N; a γ phase with an fcc arrangement of Mo atoms with N random on half the sites at a composition near Mo_2N; and a δ-MoN with the WC type of structure (see Fig. 17).

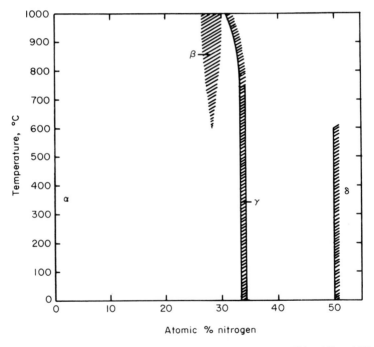

Fig. 17. Proposed phase diagram for molybdenum–nitrogen. [After Hägg *(66)*.]

The β phase is stable only above 600°C. The c/a ratio of this phase is very close to 1 ($a = 4.180$ Å, $c = 4.016$ Å) and increases toward 1 as N is added to the structure. Hägg suggested that, at temperatures higher than 600°C, β may join continuously with γ-Mo_2N, which has a lattice parameter of $a = 4.128$ Å.

Evans and Jack (67) disagree with Hägg about the Mo–N diagram. They propose that β-$MoN_{0.4}$ is a low-temperature, ordered, tetragonal modification of the face-centered γ phase ($MoN_{0.38}$–$MoN_{0.43}$; a = 4.137–4.157 Å). The γ phase is retained only by quenching from above 700°C. Furthermore, the face-centered tetragonal phase attributed to β by Hägg is, according to these investigators, only a pseudocell. The superstructure has $a' = a_{Hägg}$ and $c' = 2c_{Hägg}$. The new cell contains 8 Mo atoms at sites $8(e)$: z = 0.258 ± 3 in the space group $I4_1/amd$. Three N atoms randomly occupy four interstitial sites at (a), and the remaining four octahedral interstices are always empty.

For δ-MoN, Hägg also reported several extra X-ray lines which do not belong to the WC pattern. He tentatively assigned the N atoms to the interstitial trigonal prism sites. When Schönberg (68) reinvestigated this system, he suggested that the WC structure is merely a subcell and that all lines are identified with a superlattice structure with $a' = 2a_{Hägg}$, $c' = 2c_{Hägg}$. The phase belongs to the D_{6h}^4 space group with a slight shift of the Mo atom positions from their corresponding simple hexagonal position.

Troitskaja and Pinsker (69–71) reinvestigated this system and used thin nitrided Mo films and electron diffraction techniques. They assigned δ-MoN (renamed δ') to the D_{3d}^3 space group and established the positions of the N atoms. The Mo atoms have the same position assigned by Schönberg. In the thin films, several new phases were also identified: δ''-MoN and $Mo_{0.82-0.85}N$.

It is well known that thin-film structures do not necessarily correspond to the crystal structures of bulk samples. Therefore, the relationship of the work by Troitskaya and Pinsker to that by Hägg and Schönberg is unclear.

(3) Tungsten–nitrogen. Tungsten does not dissolve appreciable amounts of nitrogen as a terminal solid solution. Two intermediate phases, β-W_2N and δ-WN, can be formed by nitriding W powders in NH_3 at temperatures less than 800°C (66, 68). These nitrides are unstable at higher temperatures, form only at very slow rates, and cannot be prepared in an N_2 atmosphere.

The phase β-W_2N has a fcc arrangement of metal atoms and a probable statistical arrangement of N atoms (66). The unit cell size is a = 4.126 Å in a sample prepared between 700–800°C (72).

A cubic γ-phase has also been reported in samples nitrided at 825–875°C in ammonia and quenched. The structure is probably related to that of β-W_2N (72).

The phase δ-WN has the WC[B_h] type of structure with a = 2.893, c = 2.826 Å (68). The exact composition of this phase is unknown, but it probably is WN.

In addition to the above phases, a number of different tungsten nitrides have been prepared by nitriding thin W films in NH_3. These phases and their crystal structures have been discussed in Chapter 2.

IV. Ternary Systems

Ternary phase diagrams of transition-metal carbides and nitrides have been extensively measured, primarily by Rudy and co-workers[1] and by Brauer and co-workers (22, 73). These diagrams are very complex functions of temperature and composition. Figure 18 illustrates some of the com-

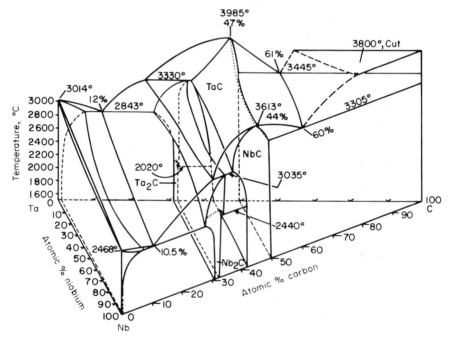

Fig. 18. Ternary phase diagram for Nb–Ta–C. [After Rudy (2).]

plexities involved in just one of these diagrams. The number of possible ternary diagrams involving the fourth, fifth, and sixth group transition metals and C and N is very large; description of even the small fraction of those that have been measured is beyond the scope of this text.

[1] E. Rudy and co-workers have published a number of ternary phase diagrams in several volumes of AFML-TR-65-2, Air Force Materials Laboratory, Research and Technology Division, Wright-Patterson A.F.B., Ohio.

Generally, these diagrams are only partially investigated over a limited composition and temperature range. Since the pseudobinary systems involving monocarbides and mononitrides have particular interest because of their high melting points and great hardness, they have been investigated extensively (*11, 12, 74, 75*). Nearly all monocarbides and mononitrides form complete solid solutions with one another. Solid-solution formation is governed by the same principles established by Hume-Rothery for the simple metals: atomic size differences of the atoms and crystal structure similarities. VC_{1-x}, for example, does not form a complete solid solution with ZrC_{1-x} because of the great disparity in the atomic sizes of V and Zr.

The degree of solid solution formation depends upon temperature and stability of certain crystal structures. MoC_{1-x} and WC_{1-x} form complete solid solutions with the fourth and fifth-group monocarbides only at the very high temperatures at which the NaCl crystal structures are stable for these sixth-group carbides.

The extensive mutual solubility of the monocarbides and the mononitrides indicates that the atomic bonding in both the carbides and the nitrides is similar. These considerations are discussed in Chapter 8.

References

1. E. K. Storms, "The Refractory Carbides." Academic Press, New York, 1967.
2. E. Rudy, "Compendium of Phase Diagram Data," AFML-TR-65-2, Part V (1969). In addition to the compendium, Rudy and co-workers have published many phase diagrams in this series.
3. E. Rudy and D. P. Harmon, AFML-TR-65-2, Part I, Vol. V (1966).
4. K. Kuo and G. Hägg, *Nature* **170**, 245 (1952).
5. H. Nowotny, E. Parthé, R. Kieffer, and F. Benesovsky, *Monatsh. Chem.* **85**, 255 (1954).
6. N. Schönberg, *Acta Chem. Scand.* **8**, 208 (1954).
7. G. Brauer and J. Jander, *Z. Anorg. Allg. Chem.* **270**, 160 (1952).
8. G. Brauer and R. Esselborn, *Z. Anorg. Allg. Chem*, **309**, 151 (1961).
9. N. Schwartz, "Vacuum Symposium Transactions," p. 325. Macmillan, New York, 1963.
10. M. Hansen, prepared with the cooperation of K. Anderko, "Constitution of Binary Alloys," 2nd ed. McGraw-Hill, New York, 1958.
10a. R. P. Elliot, "Constitution of Binary Alloys, First Supplement." McGraw-Hill, New York, 1965.
11. P. Schwartzkopf and R. Kieffer, in collaboration with W. Leszynski and F. Benesovsky, "Refractory Hard Metals." Macmillan, New York, 1953.
12. R. Kieffer and F. Benesovsky, "Hartstoffe." Springer, Vienna, 1963.
13. E. K. Storms, LAMS-2674, Part II (1962).
14. G. V. Samsonov and Ya. S. Umanskiy, "Tverdyye Soyedineniya Tugoplavkikh

Metallov." State Sci.-Tech. Lit. Publ. House, Moscow, 1957; for English translation, see *NASA Tech. Trans.* **F-102** (1962).

15. R. V. Sara, C. E. Lowell, and R. T. Dolloff, WADD-TR-60-143, Vol. IV (1963).
16. R. V. Sara, *J. Amer. Ceram. Soc.* **48**, 243 (1965).
17. E. Rudy, D. P. Harmon, and C. E. Brukl, AFML-TR-65-2, Part I, Vol. II (1965).
18. E. Rudy, AFML-TR-65-2, Part I, Vol. IV (1965).
19 E. Rudy, S. Windisch, and C. E. Brukl, *Planseeber. Pulvermet.* **16**, 3 (1968).
20. E. K. Storms and R. J. McNeal, *J. Phys. Chem.* **66**, 1401 (1962).
21. L. M. Adelsberg and L. H. Cadoff, *J. Am. Ceram. Soc.* **51**, 213 (1968).
22. G. Brauer and R. Lesser, *Z. Metallk.* **50**, 8 (1959).
23. E. Rudy and D. P. Harmon, AFML-TR-65-2, Part I, Vol. V (1966).
24. E. Storms, B. Calken, and A. Yencha, *J. High-Temp. Sci.* (1970) to be published.
25. D. S. Bloom and N. J. Grant, *J. Metals* **2**, 41 (1950).
26. E. Rudy, S. Windisch, A. J. Stosick, and J. R. Hoffman, *Trans. AIME* **239**, 1247 (1967).
27. E. Rudy, S. Windisch, and J. R. Hoffman, AFML-TR-65-2, Part I, Vol. VI (1966).
28. R. T. Doloff and R. V. Sara, WADD-TR-60-143, Part II (1961).
29. R. V. Sara, *J. Am. Ceram. Soc.* **48**, 251 (1965).
30. A. L. Bowman, T. C. Wallace, and G. P. Arnold, *8th Int. Crystallogr. Abst.*, 1969, p. S. 226; and E. K. Storms, private communication (1969).
31. R. Juza, A. Gabel, H. Rabenau, and W. Klose, *Z. Anorg. Allg. Chem.* **329**, 136 (1964).
32. A. Brager, *Acta Physiocochim. URSS* **11**, 617 (1939).
33. H. Bell, Y. M. Shy, D. E. Anderson, and L. E. Toth, *J. Appl. Phys.* **39**, 2797 (1968).
34. H. Bell, M.S. Thesis, University of Minnesota, 1966.
35. J. Sosniak, *J. Vac. Sci. Technol.* **4**, 87 (1967); see D. S. Campbell, *Thin Solid Films* **1**, 71 (1967).
36. A. E. Palty, H. Margolin, and J. P. Nielsen, *Trans. Amer. Soc. Metals* **46**, 312 (1954).
37. P. Ehrlich, *Z. Anorg. Allg. Chem.* **259**, 1 (1949).
38. H. Nowotny, F. Benesovsky, C. Brukl, and O. Schob, *Monatsh. Chem.* **92**, 403 (1961).
39. B. Holmberg, *Acta Chem. Scand.* **16**, 1255 (1962).
40. E. Ye. Vainshtein, T. S. Verkhoglyadova, Ye. A. Zhurakovskii, and G. V. Samsonov, *Phys. Metals Metallogr. (USSR)* **12** (3), 52 (1961).
41. E. Friederich and L. Sittig, *Z. Anorg. Allg. Chem.* **143**, 293 (1925).
42. C. Agte and K. Moers, *Z. Anorg. Allg. Chem.* **198**, 233 (1931).
43. E. K. Storms, LAMS-2674, Part II (1962).
44. C. P. Wang, M.S. Thesis, University of Minnesota, 1966.
45. L. E. Toth, AFOSR-68-0265 (1968); Defense Doc. No. AD-671-944.
46. E. Rudy and F. Benesovsky, *Monatsh. Chem.* **92**, 415 (1961).
47. C. F. Lee and L. E. Toth, unpublished results, University of Minnesota, 1968.
48. A. L. Giorgi, E. G. Szklarz, and T. C. Wallace, *Brit. Ceram. Soc.* **10**, 183 (1968).
49. G. Brauer and W. D. Schnell, *J. Less-Common Metals* **6**, 326 (1964).
50. H. Hahn, *Z. Anorg. Allg. Chem.* **258**, 58 (1949).
51. L. E. Toth, C. P. Wang, and C. M. Yen, *Acta Met.* **14**, 1403 (1966).
52. K. Yvon, H. Nowotny, and R. Kieffer, *Monatsh. Chem.* **98**, 34 (1967).
53. G. Brauer and E. Esselborn, *Z. Anorg. Allg. Chem.* **309**, 151 (1951).
54. R. W. Guard, J. W. Savage, D. G. Swarthout, *Trans. AIME* **239**, 643 (1967).
55. G. Brauer and J. Jander, *Z. Anorg. Allg. Chem.* **270**, 160 (1952).
56. N. Schönberg, *Acta Chem. Scand.* **8**, 208 (1954).
57. G. Brauer and H. Kirner, *Z. Anorg. Allg. Chem.* **328**, 34 (1964).
58. N. Schönberg, *Acta Chem. Scand.* **8**, 199 (1954).

59. G. Brauer and K. H. Zapp, *Z. Anorg. Allg. Chem.* **277**, 129 (1954).
60. A. Joly, *Bull. Soc. Chim. Fr.* [2] **25**, 504 (1876); C. R. *Acad. Sci.* **82**, 1195 (1876).
61. G. Brauer, J. Weidlein, and J. Strähle, *Z. Anorg. Allg. Chem.* **348**, 298 (1966).
62. S. Eriksson, *Jernkontorets Ann.* **118**, 530 (1934).
63. Z. G. Pinsker and L. N. Abrosimova, *Sov. Phys.—Crystallogr.* **3**, 285 (1958).
64. R. Blix, *Z. Phys. Chem.* **B3**, 229 (1929).
65. N. Schönberg, *Acta Chem. Scand.* **8**, 213 (1954).
66. G. Hägg, *Z. Phys. Chem., Abt.* **B7**, 339 (1930).
67. D. A. Evans and K. H. Jack, *Acta Cryst.* **10**, 833 (1957).
68. N. Schönberg, *Acta Chem. Scand.* **8**, 204 (1954).
69. N. V. Troitskay. and Z. G. Pinsker, *Sov. Phys.—Crystallogr.* **4**, 33 (1960).
70. N. V. Troitskaya and Z. G. Pinsker, *Sov. Phys.—Crystallogr.* **6**, 34 (1961).
71. N. V. Troitskaya and Z. G. Pinsker, *Sov. Phys.—Crystallogr.* **8**. 441 (1964).
72. R. Kiessling and Y. H. Liu, *J. Metals* **3**, 639 (1951).
73. G. Brauer and W. D. Schnell, *J. Less-Common Metals* **7**, 23 (1964).
74. P. Duwez and F. Odell, *J. Electrochem. Soc.* **97**, 299 (1950).
75. H. Nowotny, F. Benesovsky, and E. Rudy, *Monatsh. Chem.* **91**, 348 (1960).

4

Thermodynamics of Refractory
Carbides and Nitrides

I. Introduction

One important use of thermodynamic data is for predicting the extent to which a substance will undergo reaction with other materials or will decompose. Since the transition-metal carbides and nitrides exist over broad composition ranges, it is necessary to know the variation of the chemical potential or activities as a function of composition in order to predict the stability of these phases in different environments. Most of the reported thermodynamic data have been determined for compositions which are stated to be very close to the stoichiometric. Variations of these values with composition are unknown, and in some cases even the exact composition of the phase whose values are reported is poorly defined. It is necessary, therefore, to discuss the techniques used in estimating the thermodynamic quantities when no experimental measurements have been performed at any composition. A number of techniques are used to estimate the chemical potentials at the nonstoichiometric compositions. The Schottky–Wagner model, which indicates qualitative trends, has been successfully applied to these systems by a number of investigators and will be discussed briefly in this chapter.

The literature on the thermodynamic properties of carbides and nitrides is extensive. A number of recent critical evaluations of these properties

102

have been prepared by Schick and co-workers (*1*), Chang (*2*), Storms (*3*), Samsonov (*4*), Kelley (*5, 6*), and Stull (JANAF) (*7*). As these authors correctly point out, it is difficult to establish reliable values, because many desired measurements have not been performed, because the measurements have been performed on poorly characterized materials, and because the measurements have not been repeated by other investigators. Frequently, there are two or more conflicting results reported in the literature.

An additional problem which causes confusion about the thermodynamic values is the large number of different critical evaluations. The evaluators are in disagreement about which data to include in the evaluation and the method of evaluating that data. In view of the significant changes in the values induced by slight changes in the metal-to-nonmetal ratio, the unknown effects of impurities, and the difficulties in characterizing the materials, it appears difficult to decide which data are pertinent and which should be discarded.

For a thermodynamic analysis of a material, three types of data are necessary: the heat capacity from 0 to 298.15°K, the heat of formation at 298.15°K, and the high-temperature heat content referred to 298.15°K or some other reference temperature. In this chapter, we rely heavily upon the recent high-temperature thermodynamic evaluations of Storms (*3*). The reason for this preference is three-fold: first, he is associated with a large research group (Los Alamos) which is responsible for much of the high-temperature data, second, his evaluations are being kept up-to-date with revisions to include new data since the publication of *The Refractory Carbides*—for several carbide systems, new free-energy-function tables are published here for the first time—and third, nonstoichiometric effects are considered in detail.

For many electrical or phonon-coupled phenomena, one is also interested in the very low-temperature heat capacities for the electronic and Debye term. We review here for the first time the low-temperature heat capacities of carbides and nitrides. The discussion of the electronic term is deferred to a later chapter. For the carbides and nitrides the Debye temperature is a poorly understood parameter; nevertheless, it is frequently used to estimate thermodynamic quantities such as standard state entropies and enthalpies. The problems inherent with such estimations are discussed in detail in hopes that some misconceptions can be clarified.

II. Low-Temperature Heat Capacities

In the very low-temperature range, the principal interests in the heat capacity measurements are the electronic specific heat coefficient γ and the

Debye temperature θ_D. The γ values are important in interpreting the electronic structure of the compounds, their bonding, and the factors important in superconductivity, but this term in the heat capacity has little significance for the higher-temperature thermodynamic properties. The Debye temperature, however, is sometimes useful in estimating enthalpies and entropies at 298.15°K. Errors in the estimation of $S^{\circ}_{298.15°K}$ become increasingly important at high temperatures because of the $T\,\Delta S$ term in the free energy.

While the evaluation of the low-temperature heat capacities for most alloys is relatively straightforward, special problems arise for the carbides and nitrides because of their very high superconducting critical temperatures which often make determinations of γ and θ_D difficult.

For nonsuperconductors heat capacities at very low temperatures can be separated into two parts, the heat capacity of the electrons C_{el} and the heat capacity of the lattice C_{lat}:

$$C_p = C_{el} + C_{lat} \cong \gamma T + aT^3 \tag{1}$$

Values of γ and a are determined by plotting C_p/T versus T^2 (Fig. 1a) which theoretically should yield a straight line. Real materials, however, deviate from the ideal Debye equation and the heat capacity is better represented by a polynomial of the form:

$$C_p = \gamma T + aT^3 + bT^5 + cT^7 + \cdots \tag{2}$$

The low-temperature heat capacity of a superconductor is illustrated in Fig. 1b. γ and θ_D for a superconductor are obtained by extrapolating the normal state below superconducting transition temperature, T_c. For many high T_c carbides and nitrides the normal state behavior above T_c is not ideal and special extrapolation procedures must be applied to obtain γ and θ_D. Some of the differences between reported γ and θ_D values by different investigators have resulted from different extrapolation procedures.

One promising extrapolation procedure has been adopted by Pessall *et al.* (8) for high T_c carbides and nitrides. The heat capacity of the normal state below T_c is assumed to be of the form:

$$C_p = \gamma T + aT^3 + bT^5 \tag{3}$$

where γ, a, and b are calculated from the following conditions:

$$(C_p(T_n))\text{measured} = \gamma T_n + aT_n^3 + bT_n^5 \tag{4}$$

$$(dC_p(T_n)/dT)\text{measured} = \gamma + 3aT_n^2 + 5bT_n^4 \tag{5}$$

$$\left(\int_0^{T_n}(C_p/T)\,dT\right)\text{measured} = \gamma T_n + (aT_n^3/3) + (bT_n^3/5) \tag{6}$$

where T_n is some temperature above T_c at which the sample can be safely assumed to be completely normal. Equation (6) results from the third law of thermodynamics and the observation that the superconducting to normal transition at T_c in zero magnetic field occurs with zero entropy change.

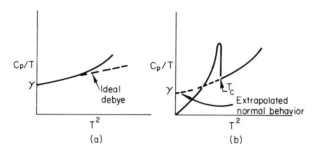

Fig. 1. The low-temperature heat capacity of a nonsuperconducting solid usually obeys an ideal Debye behavior (a) at sufficiently low temperatures. For superconductors (b) there is a jump in the heat capacity at the critical temperature T_c. For carbides and nitrides, which often have very high T_c's, the estimation of γ and θ_D is difficult because a long extrapolation must be made to $T = 0$ and because the solid may not obey an ideal Debye behavior above T_c.

Table I lists the γ and θ_D values for carbides and nitrides along with their crystal structure types and lattice parameters. All values are given on the basis of 1 gm-transition metal atom. There is generally good agreement among different investigators on the γ value for a particular carbide or nitride, but there is less agreement for the θ_D values. The discrepancies arise partly from the small lattice contributions to the heat capacity, from the inherent difficulties of accurate measurements, and partly, in the case of high T_c superconductors, from different extrapolation procedures. In Table I only the results of Pessall et al. (8) and Toth et al. (9–11) have been analyzed by the procedure outlined above.

Values of γ and θ_D have been determined for nearly all stoichiometric fourth to sixth-group carbides and nitrides with the NaCl-type crystal structure. Several monocarbides and nitrides have also been investigated as a function of composition. Ternary solid solutions between NaCl-structured carbides and nitrides generally have not been investigated nor have carbides and nitrides with non-NaCl-type structures.

TABLE I

LOW TEMPERATURE SPECIFIC HEATS OF CARBIDES AND NITRIDES

Alloy system	Structure	Lattice parameter (Å)	$\gamma(\text{mjK}^{-2}\,\text{gm-m}^{-1})$	$\theta_D(°\text{K})$	Ref.
ScC	$B1$	—	6	—	(a)
TiC	$B1$	$a = 4.329$	0.51	—	(b)
	$B1$	—	0.75	614	(c)
$TiC_{0.955}$	$B1$	$a = 4.327$	1.02	676	(d)
ZrC	$B1$	—	0.75	491	(c)
HfC	$B1$	$a = 4.636$	0.75	436	(d)
$VC_{0.88}$	$B1$	—	3.15	659	(c)
$VC_{0.84}$	$B1$	—	3.0	—	(e)
$VC_{0.83}$	$B1$	—	2.8	466	(c)
$VC_{1/2}$	Orthorhombic	$a = 2.873$ $b = 10.250$ $c = 4.572$	2.26	490	(f)
$V_{1/2}Mo_{1/2}C$	Hexagonal	$a = 2.9535$	3.16	468	(f)
$NbC_{0.98}$	$B1$	$a = 4.470$	2.64	464	(g)
$NbC_{0.98}$	$B1$	—	2.83	546	(c)
$NbC_{0.98}$	$B1$	$a = 4.469$	2.83	604	(h)
$NbC_{0.95}$	$B1$	$a = 4.467$	2.59	492	(g)
$NbC_{0.91}$	$B1$	$a = 4.466$	2.52	555	(h)
$NbC_{0.86}$	$B1$	$a = 4.461$	2.22	542	(h)
$NbC_{0.83}$	$B1$	$a = 4.453$	2.15	521	(h)
$NbC_{0.77}$	$B1$	$a = 4.448$	2.11	500	(h)
$NbC_{0.75}$	$B1$	$a = 4.447\text{--}4.436$	2.09	500	(g)
$NbC_{1/2}$	Orthorhombic	$a = 3.0955$ $b = 10.904$ $c = 4.967$	0.839	662	(f)
$NbC_{0.48}$	$\epsilon\text{-Fe}_2N$	$a = 5.402$ $c = 4.959$	1.57	464	(h)
TaC	$B1$	$a = 4.4530$	3.2	—	(b)
TaC	—	—	2.80	489	(c)
$TaC_{0.95}$	$B1$	$a = 4.450$	2.87	489	(h)
$TaC_{0.93}$	$B1$	$a = 4.447$	2.68	483	(h)
$TaC_{0.83}$	$B1$	$a = 4.430$	2.11	434	(h)
$TaC_{0.78}$	$B1$	$a = 4.423$	2.05	418	(h)
$TaC_{0.47}$	C6	$a = 3.104^a$ $c = 4.940$	1.20	378	(h)
$\alpha\text{-}MoC_{0.69}$	$B1$	$a = 4.281$	4.40	620	(i)
$\eta\text{-}MoC_{0.64}$	Hexagonal	$a = 3.010$ $c = 14.62$	3.79	536	(i)
$\alpha\text{-}MoC_{0.54}$	Orthorhombic	$a = 4.736$ $b = 6.024$ $c = 5.217$	3.41	473	(i)
$\beta\text{-}MoC_{1/2}$	Hexagonal	$a = 3.007$ $c = 4.729$	2.93	531	(f)

TABLE I *(cont.)*

Alloy system	Structure	Lattice parameter (Å)	γ(mjK^{-2} gm-m^{-1})	θ_D(°K)	Ref.
β-MoC$_{1/2}$		$a = 2.995$ $c = 4.730$	2.94	492	(i)
MoC$_{1/2}$B$_{1/2}$	Orthorhombic	$a = 3.086$ $b = 17.35$ $c = 3.047$	4.25	536	(i)
WC	Hexagonal	$a = 2.907$ $c = 2.837$	0.79	493	(d)
LaN	B1	—	3.5	300	(j)
ZrN	B1	—	2.67	515	(c)
TiN	B1	—	2.5	—	(a)
		—	3.3	636	(c)
HfN	B1	—	2.73	421	(c)
VN	B1	—	4.5	—	(a)
			8.6	420	(c)
NbN$_{0.91}$	B1	$a = 4.391$	2.64	307	(g)
NbN$_{0.84}$	B1	$a = 4.383$	3.01	331	(g)
NbN$_x$	B1	—	4.08–4.56	363–405	(c)
NbN$_{0.81}$C$_{0.09}$	B1	—	3.44	323	(c)
NbN$_{0.73}$C$_{0.17}$	B1	—	4.34	347	(c)
NbN$_{0.63}$C$_{0.27}$	B1	—	4.59	371	(c)

[a] These parameters correspond to indexing the structure as L_3' type.

References for Table I

a. P. Costa, PhD, Thesis, Orsay, 1964, reported by P. Costa, in "Anisotropy in Single-Crystal Refractory Compounds" (F. W. Vahldiek and S. A. Mersol, eds.), p. 151. Plenum Press, New York, 1968.

b. P. Costa and R. R. Conte, in "Compounds of Interest in Nuclear Reactor Technology" (J. T. Waber, P. Chiotti, and W. N. Miner, eds.), Inst. Metals Div., Spec. Rep. No. 13, p. 29. Edwards, Ann Arbor, Michigan, 1964.

c. N. Pessall, J. K. Hulm, and M. S. Walker, Final Rep., Westinghouse Research Laboratories, AF 33 (615)-2729 (1967).

d. Y. A. Chang, L. E. Toth, and Y. S. Tyan, unpublished results (1969).

e. Unpublished research by Bonnerot, Orsay (reported in ref. a).

f. R. Caudron, P. Costa, and B. Sulgeot, *2nd Int. Conf. Semi-Met. Compounds Transition Elements ENSCHEDE, 1967* reported by P. Costa, *in* "Anisotropy in Single-Crystal Refractory Compounds" (F. W. Vahldiek and S. A. Mersol, eds.), p. 151. Plenum Press, New York, 1968.

g. T. H. Geballe, B. T. Matthias, J. P. Remeika, A. M. Clogston, V. B. Compton, J. P. Maita, and H. J. Williams, *Physics (Long Island City, N.Y.)* 2, 293 (1966).

h. L. E. Toth, M. Ishikawa, and Y. A. Chang, *Acta Met.* 16, 1183 (1968).

i. L. E. Toth, J. Zbasnik, Y. Sato, and W. Gardner, *in* "Anisropy in Single-Crystal Refractory Compounds" (F. W. Vahldiek and S. A. Mersol, eds.), p. 249 Plenum Press, New York, 1968.

j. L. E. Toth and J. Zbasnik, *Acta Met.* 16, 1177 (1968).

Since γ values yield information about the density of electron states at the Fermi level, we defer their discussion to subsequent chapters on electrical properties and bonding.

Debye temperatures for carbides and nitrides are determined from the coefficient a in Eqs. (1) and (2) by the expression

$$\theta_D = (234 N_0 k_B / a)^{1/3} \tag{7}$$

where N_0 is Avagadro's number and a is based on 1 gm-transition metal atom. Using this definition of θ_D really assumes that the principal phonon contribution at very low temperatures is due to the transition metal vibrations.

It is interesting to note that θ_D sharply increases in the monocarbides NbC_{1-x} and TaC_{1-x} as the C/Me ratio is increased. This behavior is shown in Fig. 2. This means that the total lattice contribution to the heat capacity

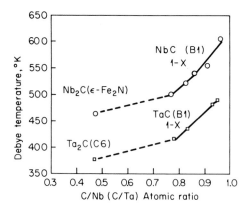

Fig. 2. The Debye temperatures of NbC_{1-x} and TaC_{1-x} are sensitive functions of the carbon-to-metal ratio. [After Toth et al., Acta Met. 16, 1183 (1968). Reprinted by permission of Pergamon Press.]

actually decreases as more carbon is dissolved in the monocarbide (decreasing a). The probable explanation for the increase in θ_D is the greatly increased bond strength between metal atoms and metal–nonmetal atoms. This increased bond strength "stiffens" the lattice, decreases the atomic vibrations and increases θ_D.

Debye temperatures can also be determined by a number of other techniques including elastic constant measurements. Generally for most elements and compounds θ_D values determined by low-temperature heat capacity measurements are in good agreement with Debye temperatures determined by elastic constant measurements, $\theta_D^{(E)}$. Chang et al. (12) compared θ_D and $\theta_D^{(E)}$ for carbides and found that the two values agree

TABLE II

COMPARISON OF θ_D VALUES DETERMINED BY ELASTIC CONSTANT AND
LOW-TEMPERATURE C_p MEASUREMENTS [a]

Carbide	$\theta_D^{(E)}(°K)$ [b] (Elastic const.)	$\theta_D^{(E)}(°K)$ [c]	$\theta_D(°K)$ [d] (C_p)	$\theta_D(°K)$ [c]	Difference (%)
TiC	946 (a), 925 (b), 950 (c)	940	774 (h), 845 (i)	845	+10.0
ZrC	691 (d), 694 (e), 714 (a)	700	619 (h), 649 (j)	649	+7.3
HfC	553 (f)	553	523 (k), 549 (i)	549	+0.7
NbC	742 (f)	742	688 (h), 761 (l)	761	−2.6
TaC	573 (f)	573	616 (h), 616 (l)	616	−7.5
MoC$_{1/2}$	892 (g)[e]	892	546 (m), 553 (n), 608 (o)	569	+36.0
WC	617 (f)	617	621 (i)	621	−0.6

[a] θ_D values are computed on the basis of 1 gm-atom of alloy. [After Chang *et al.* (*12*)].

[b] $\theta_D^{(E)}$ values were computed by the formula $\theta_D^{(E)} = h/k[3qN_0\rho/4\pi]_{v_m}^{1/3}$ where q is the number of atoms in the formula, ρ is the density in gm-atom of an alloy per unit volume and v_m is the mean velocity of sound, which was evaluated from single-crystal elastic-constant data by the method of O. L. Anderson [*J. Phys. Chem. Solids* **24**, 909 (1963)].

[c] Either average or preferred value.

[d] For high T_c superconductors ($>8°K$) only those θ_D values analyzed by the method of Pessall *et al.* (*8*) described in the text were included.

[e] $\theta_D^{(E)}$ value was taken from the original article.

References for Table II

a. R. Chang and L. J. Graham, *J. Appl. Phys.* **37**, 3778 (1966).
b. J. J. Gilman and B. W. Roberts, *J. Appl. Phys.* **32**, 1405 (1961).
c. B. T. Bernstein, unpublished data, reported by Chang and Graham (ref. a).
d. R. Lowrie, NP-11311 (1961), as quoted by H. L. Brown and C. P. Kempter (see ref. e).
e. H. L. Brown and C. P. Kempter, *Phys. Status Solidi* **18**, K21 (1966).
f. H. L. Brown, P. E. Armstrong, and C. P. Kempter, *J. Chem. Phys.* **45**, 547 (1966).
g. F. W. Vahldiek, S. A. Mersol, and C. T. Lynch, Tech. Rep. AFML-TR-66-28 (1966).
h. N. Pessall, J. K. Hulm, and M. S. Walker, Final Rep., Westinghouse Research Laboratories, AF-33(615)-2729 (1967).
i. Y. A. Chang, L. E. Toth, and Y. S. Tyan, *Trans. AIME*, to be published.
j. E. F. Westrum, Jr. and G. Feick, *J. Chem. Eng. Data* **8**, 176 (1963).
k. E. F. Westrum, Jr., ASD-TDR-62-204, Part III (1964).
l. L. E. Toth, M. Ishikawa, and Y. A. Chang, *Acta Met.* **6**, 1183 (1968).
m. L. E. Toth, J. Zbasnik, W. Gardner, and Y. Sato, in "Anisotropy in Single-Crystal Refractory Compounds" (F. W. Vahldiek and S. A. Mersol, eds.), Vol. 1, p. 249. Plenum Press, New York, 1968.
n. L. E. Toth and J. Zbasnik, *Acta Met.* **16**, 1177 (1968).
o. R. Caudron, P. Costa, and B. Sulgeot, *2nd Int. Conf. Semi-Met. Compounds Transition Elements ENSCHEDE, 1967*, reported by P. Costa *in* "Anisotropy Single-Crystal Refractory Compounds" (F. W. Vahldiek and S. A. Mersol, eds.), p. 151. Plenum Press, New York, 1968.

for most phases to within 10% which is a reasonable estimate of the uncertainty in θ_D values. For Mo_2C and possibly TiC a more serious discrepancy exists between the two values. This comparison is shown in Table II (12). To prepare Table II, the θ_D values listed in Table I were recomputed to a value corresponding to 1 gm-atom of alloy since $\theta_D^{(E)}$ values are usually computed on this basis. Several reasons could account for part or all of the differences in the two values for Mo_2C and TiC. These reasons include:

(1) Experimental determinations of θ_D by low temperature heat capacities are difficult due to the relatively small lattice contributions to the heat capacities [small a values in Eq. (7)] and sometimes due to the high superconducting critical temperatures; these difficulties lead to an estimated error in θ_D of about $\pm 10\%$ between measurements by different research groups.

(2) The temperature of elastic constant measurement for Mo_2C was not 4.2°K and $\theta_D^{(E)}$ is known to vary with temperature.

(3) Impurities such as TiO in TiC could affect θ_D values because of the large difference in Debye values for these impurity phases; when measured chemically, however, the oxygen content was small in samples used in both types of measurements.

(4) Different C/Me ratios were involved in the two experiments and the ratios were not known accurately because of poor chemical analysis.

While these reasons could account for the difference in Debye values for TiC, it is doubtful if the discrepancy for Mo_2C can be explained on this basis. For this reason it is suggested that both the low-temperature heat capacities and elastic constant measurements on Mo_2C be repeated.

It is important to point out that θ_D values derived from Eq. (7) should not be used to estimate entropy and enthalpy values at 298°K. Even at low temperatures, the heat capacities of carbides and nitrides deviate considerably from the ideal Debye behavior and T^5 and T^7 terms become important. As shown in the next section, however, some investigators have estimated entropies and enthalpies at 298°K by using a single θ_D value estimated from melting points which, coincidentally, turns out to be in fair agreement with the low temperature heat capacity θ_D. As expected, very little agreement is reached with the experimental entropy and enthalpy data.

III. Standard State Entropies and Enthalpies at 298.15°K

Standard state entropies $S^\circ_{298.15°K}$ and enthalpies $H^\circ_{298.15°K}$ can be calculated from heat capacity data by the formulas:

$$S^{\circ}_{298.15°K} = S^{\circ}_{0°K} + \int_0^{298.15} (C_D/T)\, dT, \tag{8}$$

and

$$H^{\circ}_{298.15°K} = H^{\circ}_{0°K} + \int_0^{298.15} C_p\, dT. \tag{9}$$

Usually the heat capacity is measured from 298 to about 50°K and then extrapolated to lower temperatures using appropriate Debye functions. For an ordered stoichiometric carbide or nitride $S^{\circ}_{0°K} = 0$, but for non-stoichiometric compositions $S^{\circ}_{0°K}$ is positive due to mixing of atoms and vacancies. For many stoichiometric carbide and nitrides vacancies exist on both sublattices and in this case $S^{\circ}_{0°K} \neq 0$.

Experimental values of $(S^{\circ}_{298.15°K} - S^{\circ}_{0°K})$ and $(H^{\circ}_{298.15°K} - H^{\circ}_{0°K})$ for carbides and nitrides are listed in Table III. Much of the original data has been critically evaluted (1–3, 6) for impurities such as free carbon, and the corrected values are given here. For the carbides, the values of Storms (3) are used so that the data will be consistent with the high temperature thermal functions presented later. The differences between the three evaluations (1–3) at 298°K are small and less than 1%. Most measurements are for compositions stated to be nearly stoichiometric; one has to remember, however, that these phases are difficult to analyze chemically and that therefore the actual compositions may be inexactly known. The data for the nitrides in Table III is particularly scarce.

Only for NbC_{1-x} have $(S^{\circ}_{298.15°K} - S^{\circ}_{0°K})$ and $(H^{\circ}_{298.15°K} - H^{\circ}_{0°K})$ been measured as a function of the nonmetal-to-metal ratio. Both $(S^{\circ}_{298.15°K} - S^{\circ}_{0°K})$ and $(H^{\circ}_{298.15°K} - H^{\circ}_{0°K})$ are strongly dependent upon composition (Table III). Presumably these parameters in other binary carbide and nitride systems would also show strong dependencies upon composition. This is one area where further research on well characterized samples is badly needed. Measurements have also not been conducted on the effect of vacancy concentrations at a fixed composition for these and other thermodynamic parameters. For the nitrides, in particular, the vacancy concentration on both sublattices can be several percent at stoichiometry. The vacancy concentration is controlled to a limited extent by a combination of nitriding pressure and temperature and therefore different sample preparation techniques could inadvertently result in different vacancy concentrations. The introduction of an appreciable quantity of vacancies (1–2%) will alter the phonon spectrum and thus $(S^{\circ}_{298.15°K} - S^{\circ}_{0°K})$ and $(H^{\circ}_{298.15°K} - H^{\circ}_{0°K})$. Certainly, for a complete thermodynamic description, knowledge of the vacancy concentration on both sublattices is necessary and greater care is needed to specify that quantity when it can be significant.

When experimental data are not available, the standard state entropies

TABLE III

EXPERIMENTAL VALUES OF $S^\circ_{298.15°K} - S^\circ_{0°K}$, $H^\circ_{298.15°K} - H^\circ_{0°K}$, AND $C_p(298.15°K)$ FOR CARBIDES AND NITRIDES [a]

Phase	Nonmetal/metal ratio	$S^\circ_{298.15°K} - S^\circ_{0°K}$ cal^{-1} °K^{-1} gm-atom Me^{-1}	$H^\circ_{298.15°K} - H^\circ_{0°K}$ kcal gm-atom Me^{-1}	$C_p(298.15°K)$ cal^{-1} °K^{-1} gm-atom Me^{-1}	Temperature range (°K) of measurement	Reference
TiC	1.0	5.79	1.102	8.08	55–295	(a)
ZrC	0.96	7.93	1.394	9.02	5–350	(b)
HfC	0.968	9.43	1.523	8.96	5–350	(c)
VC	0.887	6.61	1.179	7.72	52–297	(d)
NbC	0.996	8.46	1.424	8.80	51–297	(e)
	0.980	8.29	1.390	8.79	7–320	(f)
	0.825	7.87	1.335	8.41	7–320	(f)
	0.702	7.63	1.280	7.97	7–320	(f)
βNb$_2$C	0.500	7.66	1.270	7.59	7–320	(f)
TaC	1.0	10.1	1.558	8.76	54–295	(g)
CrC$_{6/23}$	6/23	6.34	1.087	6.53	54–295	(h)
CrC$_{3/7}$	3/7	6.86	1.174	7.14	53–296	(h)
CrC$_{2/3}$	2/3	6.81	1.207	7.84	12–301	(h, i)
TiN	1.0	7.24	1.310	8.86	52–297	(j)
ZrN	1.0	9.29	1.575	9.65	5–297	(k, l)
VN	1.0	8.91	—	9.08	—	(m)

[a] Evaluations for carbides are primarily from Storms (3).

and enthalpies are roughly estimated by using the Lindemann formula and the melting point to calculate a Debye temperature. The heat capacities at constant volume are then calculated and converted to constant pressure heat capacities with the Nernst–Lindemann (13) relationship. The method which apparently yields the best agreement with experimental values for entropy and enthalpy is the one in which a θ_D is calculated separately for each element (the metal and nonmetal atom) in the compound by the relationship (1)

$$\theta_{D_i} = K(T_m/M_i)^{1/2} V_i^{-1/3}. \tag{10}$$

Here the subscript i refers to the metal or nonmetal atom in the formula, K is a constant about 137, T_m is the melting point of the compound, M_i and V_i are the molecular weight and volume of the ith element.

Table IV compares several calculated entropy values at 298.15°K with experimental values. For most carbides, this method yields reasonable but not accurate values. For the nitrides, the limited comparison indicates poorer agreement, probably because of greater uncertainty in estimating the melting points.

Kaufman (14) has proposed a slightly different method for estimating entropies and enthalpies using a single θ_D value which is also estimated from the melting point of the compound. As seen from Table IV, Kaufman's method yields very poor agreement with the experimental values. Kaufman's θ_D values, however, are in reasonable agreement with the experimental θ_D values derived from low temperature heat capacities and, therefore, these latter values should also not be used in estimating entropy values.

One problem in estimating entropies and enthalpies by these methods

---

References for Table III

a. K. K. Kelley, *Ind. Eng. Chem.* **36**, 805 (1944).
b. E. F. Westrum, Jr. and G. Feick, *J. Chem. Eng. Data* **8**, 176 (1963).
c. L. A. McClaine, Wright-Patterson Air Force Base, Tech. Rep. ASD-TDR-62-204, Part III (1964). (Data by Westrum.)
d. C. H. Shomate and K. K. Kelley, *J. Amer. Chem. Soc.* **71**, 314 (1949).
e. L. B. Pankratz, W. W. Weller, and K. K. Kelley, *U.S. Bur. Mines Rep. Invest.* RI-**6446** (1964).
f. T. A. Sandenaw and E. K. Storms, Los Alamos Scientific Laboratory, LA-3331 (1965), *J. Phys. Chem. Solids* **27**, 217 (1966).
g. K. K. Kelley, *J. Amer Chem. Soc.* **62**, 818 (1940).
h. K. K. Kelley, F. S. Boericke, G. E. Moore, E. H. Huffman, and W. M. Bangert, *U.S. Bur. Mines, Tech. Pap.* **662** (1949).
i. W. DeSorbo, *J. Amer. Chem. Soc.* **75**, 1825 (1953).
j. C. H. Shomate, *J. Amer. Chem. Soc.* **68**, 310 (1946).
k. S. S. Todd, *J. Amer. Chem. Soc.* **72**, 2914 (1950).
l. K. K. Kelley and E. G. King, *U.S. Bur. Mines, Bull.* **592** (1961).
m. C. H. Shomate and K. K. Kelley, *J. Amer. Chem. Soc.* **71**, 314 (1949).

TABLE IV

COMPARISON OF ESTIMATED AND EXPERIMENTAL
$S^\circ_{298.15^\circ K} - S^\circ_{0^\circ K}$ VALUES [a]

Compound	Calculated values		Experimental values from Table III
	Schick (1)	Kaufman (14)	
TiC	6.18	7.1	5.79
ZrC	8.11	10.80	7.93
HfC	9.85		9.43
VC	6.44	7.5	6.61
V_2C	5.93		
NbC	7.41	11.4	8.29
Nb_2C	7.59		7.66
TaC	9.29		10.1
Ta_2C	9.07		
TiN	7.89		7.24
ZrN	10.12		9.29
HfN	11.49		
Nb_2N	9.17		
NbN	11.39		
TaN	11.41		
Ta_2N	10.62		

[a] All values are listed in cal $^\circ K^{-1}$ gm-atom Me^{-1}.

relying on melting points is that there is no means for calculating these quantities as a function of metal-to-nonmetal ratio. As seen in Table III the entropy value for NbC_{1-x} varies by nearly 15% from $NbC_{1.0}$ to $NbC_{0.7}$. Yet, over this composition range there is only one point at which congruent melting occurs and thus a unique melting point, in general, cannot be defined for an arbitrary composition. This observation plus the empiricism of the approach emphasize the need for more experimental determinations of $S^\circ_{298.15^\circ K} - S^\circ_{0^\circ K}$ and $H^\circ_{298.15^\circ K} - H^\circ_{0^\circ K}$ on well characterized specimens particularly as a function of nonmetal-to-metal ratio.

Use of Debye temperatures is not limited to the estimation of standard state entropies and enthalpies at 298°K. Chang (15), for example, fitted experimental C_p data for several carbides in the temperature range 325–985°K to an empirical function involving two Debye temperatures—one for the metal and one for the nonmetal. Using these empirically determined Debye temperatures he then extrapolates C_p and related thermodynamic parameters to the respective melting points. As a check of the method, he finds good agreement between calculated and experimental of the linear coefficient of thermal expansion.

In reviewing the use of estimation techniques involving Debye temperatures we find cases in which two temperatures are estimated from equations like Eq. (10) and cases in which experimental data are fitted over a temperature region to a function involving two empirically determined Debye temperatures. It should be emphasized that two Debye temperatures have no physical meaning and that the values are not related to the single value derived from low temperature C_p and elastic constant data. Use of two adjustable θ_D's, however, allows an additional parameter in fitting known high temperature C_p data and one can speculate that the relative success of this method over the single θ_D method is solely due to the extra degree of freedom in curve fitting allowed by the additional parameter. In evaluating the usefulness of these techniques we should also keep clearly in mind the distinction between methods in which θ_D's are estimated by equations like Eq. (10) and methods in which θ_D's are empirically derived by curve fitting.

IV. High-Temperature Thermodynamic Properties

In addition to standard-state entropies and enthalpies, a thermochemical description of a material requires knowledge of the heat of formation at 298.15°K and the heat content at the temperature of interest referred to 298.15°K. Heats of formation can be determined with a reaction calorimeter; the carbide or nitride is converted to an oxide and the evolved heat measured. Heat contents are usually measured with a drop calorimeter. For carbides and nitrides, these parameters vary significantly with deviations from stoichiometry and the variations are difficult to estimate with any degree of accuracy. To experimentally measure these parameters as a function of composition, however, is a difficult task because of the very high temperatures involved and problems of controlling composition. Most measurements have been confined to compositions closest to stoichiometry. For a few carbides heats of formation and heat contents have been studied as a function of composition.

IVA. Heats of Formation

Figures 3–5 show how heats of formation for Hf–C, Nb–C, and Ta–C vary with composition (3, 16, 29). The figures are from Storms (3). The reader is cautioned about the scatter in the data. An individual measurement at a given composition is probably no better than ±10%. This scatter illustrates the need for further careful experiments on the thermodynamic properties as a function of composition. Sources of error in the measurements include uncertainties in the C/Me or N/Me ratio, the amount of impurities and the chemical form of the impurities. Even in careful work

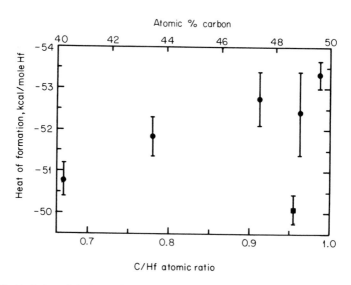

Fig. 3. Variation of the heat of formation in HfC$_{1-x}$ with carbon content. ● : data from Zhelankin and Kutsev (*29*) ; ■ : data from Mah (*16*). [After Storms (*3*).]

the uncertainty in carbon or nitrogen content may be a few atomic percent and in less careful work a 3–5% uncertainty is not uncommon. Particularly for nitrides the larger error limits may apply. The uncertainties in determining composition must be combined with those due to difficulties in performing high temperature measurements. Thus, it is not surprising to find different research groups reporting experimental values for the supposedly same composition which differ by 10% or more.

Further values of heats of formation for carbides and nitrides are listed in Table V. These values are based on still less extensive measurements than those for NbC$_{1-x}$ and TaC$_{1-x}$ and are therefore subject to greater uncertainty. The values listed are those preferred in the evaluations of Storms (*3, 30*) and Schick (*1*). For carbides, Storms has given a more nearly complete evaluation and listing of the experimental values. The reader should consult his book for a better understanding of the uncertainties involved for each composition.

Figure 6 shows the heats of formation at 298.15°K for nearly stoichiometric carbides as a function of the group number of the transition metal and Fig. 7 presents similar information for the nitrides. Within each group, the heats of formation are fairly similar, and the absolute magnitude decreases from the fourth group monocarbides and nitrides to the sixth group. For the nitrides the variation is similar to that for the melting points or the temperature of dissociation. A similar correlation exists for the

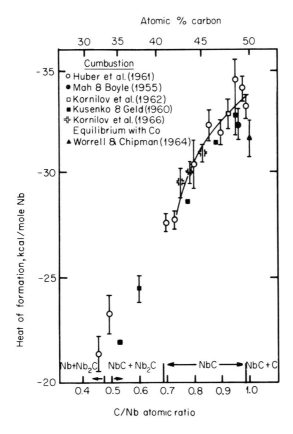

Fig. 4. Variation of the heat of formation in NbC_{1-x} with carbon content. [After Storms (3).]

carbides, although here the maximum in melting points occurs between the fourth and fifth group.

The systematic increase in the heats of formation with increasing group number has been interpreted as suggesting that the bonding portions of the electron band structure are filled at the fourth group and that higher groups become relatively less stable due to a filling of the antibonding portions of the band (31). These theories are discussed in detail in Chapter 8.

IVB. Heat Content and Heat Capacity

High temperature thermal functions are generally available only for nearly stoichiometric monocarbides. Only for NbC_{1-x} is there any information

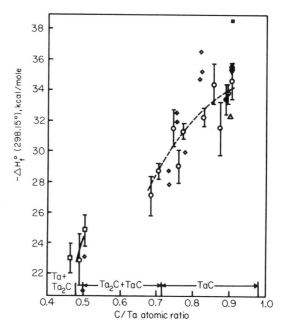

Fig. 5. Variation of the heat of formation in TaC_{1-x} with carbon content. $\bigcirc$: Huber *et al.* (*23*); $\square$: Kornilov *et al.* (*24*); $\bullet$: Kornilov *et al.* (*19*); $\triangle$: McKenna (*25*); $\blacksquare$: Humphrey (*26*); $\triangledown$: Mah (*27*); $\diamondsuit$: Smirnova and Ormont (*28*); $\blacktriangle$: Worrell and Chipman (*22*). Based on $\triangle H_{f}°(Ta_2O_5) = -488.7 \pm 0.4$ kcal/mole. [After Storms (*3*).]

TABLE V

STANDARD HEATS OF FORMATION OF CARBIDES AND NITRIDES
AT 298.15°K [a]

Phase	$-\triangle H_{f,\,\text{ST}}$ (cal/gm-atom Me)	Phase	$-\triangle H_{f,\,\text{ST}}$ (cal/gm-atom Me)
TiC	44,100+ 0, −1000	MoC	3000±2000
$ZrC_{0.93}$	47,000±600	α-$WC_{1/2}$	6300±600
$HfC_{0.958}$	50,080±350	WC	9670±400
$VC_{1/2}$	16,500±600	TiN	80,750
$VC_{0.88}$	24,500±600	ZrN	87,300
β-$NbC_{1/2}$	23,300±600	HfN	88,240
NbC	33,600±500	NbN	56,500
$TaC_{1/2}$	24,900+0, −1000	Nb_2N	30,250
TaC	34,100±500	TaN	59,950
$CrC_{2/3}$	5500±1000	Ta_2N	32,300
$MoC_{1/2}$	5500±300		

[a] Data for carbides evaluated by Storms (*3*), data for nitrides evaluated by Schick *et al.* (*1*).

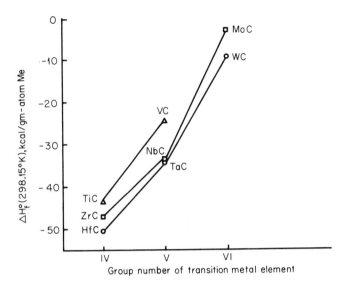

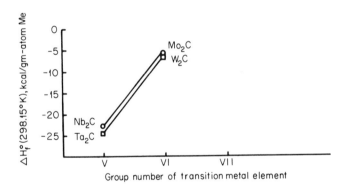

Fig. 6. Heats of formation of carbides by group number.

about the effects of composition variations. The information for the nitrides is scarce; many of the high temperature functions are not available and only estimations exist. Schick and co-workers (*1*) have compiled these functions for the nitrides.

For the carbides the experimental situation is much better than for the nitrides. The high temperature thermodynamic properties of group IV, V, and VI carbides have been evaluated by Storms (*3*), Kelley (*6*), and Chang

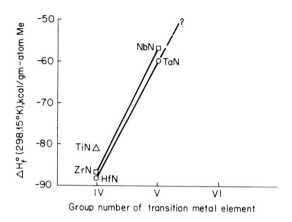

Fig. 7. Heats of formation of nitrides by group number.

(2). The most up-to-date tables of thermal functions are those of Storms (3, 32). Here we reproduce many of the thermodynamical tables which appeared in *The Refractory Carbides* (see Tables VI–XIX). Several of the tables have been updated by Storms to include new experimental data and these newer tables are included here. Another reason for preferring the Storms evaluation is that stoichiometry effects are carefully considered.

Storms' method of evaluation differs somewhat from that used by Kelley (5) and Chang (2). The usual method for treating high-temperature enthalpy data is to perform a least squares analysis of the data to the equation

$$H_T^\circ - H_{298.15^\circ K}^\circ = A + BT + CT^2 + (D/T)$$

as suggested by Maier and Kelley (33) and modified by Shomate (34). When this equation is differentiated the expression for the heat capacity basically increases linearly with temperature at high temperatures. The heat capacity for many carbides, however, increases more rapidly with temperature near the melting point. To make the C_p and $H_T^\circ - H_{298.15^\circ K}^\circ$ fit the experimental data better, Storms expanded the expression to include a T^3 term. The data evaluated in Tables VI–XIX are fitted to this expanded form of the equation. In the least squares fit the equation is set equal to zero at 298.15°K and the C_p is set equal to the experimental value of 298.15°K. Chang (2) also observed that the Kelley equation inadequately fitted the data. His approach was to split the temperature region of the fit into separate regions and obtain expressions for each section. Godfrey and Leitnaker (35) suggest the use of an additional $T^{2.5}$ term which also appears to give a satisfactory fit.

TABLE VI

THERMAL FUNCTIONS OF $TiC_{\sim 1.0}$ [a, b]

T, (°K)	$H_T^\circ - H_{298}^\circ$ (cal/mole)	C_p° (cal/mole-deg)	S_T° (cal/mole-deg)	$-(F_T^\circ - H_{298}^\circ)/T$ (cal/mole-deg)
298.15	0.00	8.080	5.790	5.790
300	15.00	8.136	5.840	5.790
400	939.7	10.07	8.487	6.138
500	1995	10.93	10.84	6.848
600	3112	11.38	12.87	7.687
700	4264	11.64	14.65	8.557
800	5437	11.82	16.22	9.419
900	6626	11.96	17.62	10.25
1000	7829	12.08	18.88	11.05
1100	9043	12.21	20.04	11.82
1200	10,270	12.33	21.11	12.55
1300	11,510	12.47	22.10	13.25
1400	12,760	12.62	23.03	13.91
1500	14,030	12.79	23.91	14.55
1600	15,320	12.97	24.74	15.16
1700	16,630	13.18	25.53	15.75
1800	17,960	13.40	26.29	16.31
1900	19,310	13.65	27.02	16.86
2000	20,690	13.92	27.73	17.38
2100	22,100	14.21	28.41	17.89
2200	23,530	14.52	29.08	18.38
2300	25,000	14.85	29.73	18.86
2400	26,500	15.20	30.37	19.33
2500	28,040	15.58	31.00	19.78
2600	29,620	15.98	31.62	20.23
2700	31,240	16.40	32.23	20.66
2800	32,900	16.84	32.83	21.09
2900	34,610	17.31	33.43	21.50
3000	36,360	17.80	34.03	21.91

[a] $H_T^\circ - H_{298.15}^\circ = -5.3007 \times 10^3 + 13.296T - 9.7189 \times 10^{-4}T^2 + 3.8451 \times 10^{-7}T^3 + 4.2124 \times 10^5/T$ (298–3000°K, cal/mole, ±0.5%) mol. wt. = 59.91.

[b] Data evaluated by Storms (3).

TABLE VII

THERMAL FUNCTIONS OF $ZrC_{0.96}$ [a, b]

T ($°K$)	$H_T^° - H_{298}^°$ (cal/mole)	$C_p^°$ (cal/mole-deg)	$S_T^°$ (cal/mole-deg)	$-(F_T^° - H_{298}^°)/T$ (cal/mole-deg)
298.15	0.00	9.016	7.927	7.927
300	16.73	9.066	7.983	7.927
400	1021	10.76	10.86	8.308
500	2136	11.45	13.35	9.074
600	3299	11.77	15.47	9.967
700	4485	11.93	17.29	10.89
800	5683	12.01	18.89	11.79
900	6887	12.07	20.31	12.66
1000	8095	12.11	21.58	13.49
1100	9309	12.16	22.74	14.28
1200	10,530	12.22	23.80	15.03
1300	11,750	12.30	24.78	15.74
1400	12,990	12.39	25.70	16.42
1500	14,230	12.51	26.56	17.07
1600	15,490	12.65	27.37	17.69
1700	16,760	12.81	28.14	18.28
1800	18,050	13.00	28.88	18.85
1900	19,360	13.22	29.59	19.39
2000	20,700	13.46	30.27	19.92
2100	22,060	13.73	30.93	20.43
2200	23,450	14.02	31.58	20.92
2300	24,860	14.34	32.21	21.40
2400	26,310	14.69	32.83	21.86
2500	27,800	15.07	33.43	22.31
2600	29,330	15.47	34.03	22.75
2700	30,900	15.90	34.62	23.18
2800	32,510	16.36	35.21	23.60
2900	34,170	16.84	35.79	24.01
3000	35,880	17.35	36.37	24.41

[a] $H_T^° - H_{298.15}^° = -5.4298 \times 10^3 + 14.228T - 1.5583 \times 10^{-3}T^2 + 4.6364 \times 10^{-7}T^3 + 3.9173 \times 10^5/T$ (298–3000°K, cal/mole, ±0.2%); mol. wt. = 102.75; $H_{298.15}^° - H_0^°$ = 1394 cal/mole.

[b] Data evaluated by Storms (3).

TABLE VIII

THERMAL FUNCTIONS OF $HfC_{0.98}$ [a, b]

T (°K)	$H_T^\circ - H_{298}^\circ$ (cal/mole)	C_p° (cal/mole-deg)	S_T° (cal/mole-deg)	$-(F_T^\circ - H_{298}^\circ)/T$ (cal/mole-deg)
298.15	0.00	8.955	9.431	9.431
300	16.60	8.986	9.486	9.431
400	979.4	10.13	12.25	9.800
500	2027	10.77	14.58	10.53
600	3126	11.19	16.59	11.38
700	4261	11.51	18.34	12.25
800	5426	11.78	19.89	13.11
900	6616	12.01	21.29	13.94
1000	7828	12.22	22.57	14.74
1100	9060	12.41	23.74	15.51
1200	10,310	12.58	24.83	16.24
1300	11,580	12.75	25.84	16.94
1400	12,860	12.90	26.79	17.61
1500	14,160	13.04	27.69	18.25
1600	15,470	13.18	28.53	18.87
1700	16,790	13.31	29.34	19.46
1800	18,130	13.43	30.10	20.03
1900	19,480	13.54	30.83	20.58
2000	20,840	13.65	31.53	21.11
2100	22,200	13.74	32.20	21.62
2200	23,580	13.84	32.84	22.12
2300	24,970	13.92	33.45	22.60
2400	26,370	14.00	34.05	23.06
2500	27,770	14.07	34.62	23.51
2600	29,180	14.14	35.17	23.95
2700	30,600	14.20	35.71	24.38
2800	32,020	14.25	36.23	24.79
2900	33,450	14.30	36.73	25.19
3000	34,880	14.34	37.21	25.59

[a] $H_T^\circ - H_{298.15}^\circ = -3.8886 \times 10^3 + 10.526T + 1.0963 \times 10^{-3}T^2 - 1.0149 \times 10^{-7}T^3 + 1.9539 \times 10^5/T$ (298–3000°K cal/mole, ±0.7%) ; mol. wt. = 190.26 ; $H_{298.15}^\circ - H_0^\circ = 1523.2$ cal/mole.
[b] Data evaluated by Storms (3).

TABLE IX

THERMAL FUNCTIONS OF $VC_{0.88}$ [a, b]

T (°K)	$H_T^\circ - H_{298}^\circ$ (cal/mole)	C_p° (cal/mole-deg)	S_T° (cal/mole-deg)	$-(F_T^\circ - H_{298}^\circ)/T$ (cal/mole-deg)
298.15	0.00	7.722	6.610	6.610
300	14.31	7.749	6.658	6.610
400	849.7	8.858	9.053	6.929
500	1774	9.581	11.11	7.565
600	2761	10.14	12.91	8.309
700	3799	10.61	14.51	9.083
800	4882	11.03	15.96	9.853
900	6004	11.41	17.28	10.61
1000	7162	11.75	18.50	11.33
1100	8352	12.06	19.63	12.04
1200	9573	12.35	20.69	12.72
1300	10,820	12.61	21.69	13.37
1400	12,100	12.86	22.64	14.00
1500	13,390	13.07	23.53	14.60
1600	14,710	13.27	24.38	15.19
1700	16,050	13.45	25.19	15.75
1800	17,400	13.60	25.96	16.30
1900	18,770	13.74	26.70	16.38
2000	20,150	13.85	27.41	17.34
2100	21,530	13.94	28.09	17.83
2200	22,930	14.02	28.74	18.31
2300	24,340	14.07	29.36	18.78
2400	25,750	14.10	29.96	19.24
2500	27,160	14.11	20.54	19.68

[a] $H_T^\circ - H_{298.15}^\circ = -3.0347 \times 10^3 + 7.8928T + 2.4967 \times 10^{-3}T^2 - 3.3282 \times 10^{-7} + 1.3964 \times 10^5/T$ (298–2500°K, ±1.0%); mol. wt. = 61.51. A randomization entropy of 0.73 eu is not included.

[b] Data evaluated by Storms (3).

TABLE X

THERMAL FUNCTIONS OF $NbC_{0.5}$ (β-Nb_2C) [a, b]

T (°K)	$H°_T - H°_{298}$ (cal/mole)	$C°_p$ (cal/mole-deg)	$S°_T$ (cal/mole-deg)	$-(F°_T - H°_{298})/T$ (cal/mole-deg)
298.15	0.00	7.590	7.660	7.660
300	14.05	7.596	7.707	7.660
400	788.2	7.884	9.932	7.962
500	1590	8.158	11.72	8.540
600	2419	8.419	13.23	9.199
700	3274	8.669	14.55	9.871
800	4153	8.907	15.72	10.53
900	5055	9.134	16.78	11.17
1000	5979	9.350	17.76	11.78
1100	6925	9.556	18.66	12.36
1200	7890	9.750	19.50	12.92
1300	8874	9.933	20.29	13.46
1400	9876	10.11	21.03	13.97
1500	10,890	10.27	21.73	14.47
1600	11,930	10.42	22.40	14.94
1700	12,980	10.56	23.04	15.40
1800	14,041	10.69	23.64	15.84
1900	15,120	10.81	24.22	16.27
2000	16,200	10.91	24.78	16.68
2100	17,300	11.01	25.32	17.08
2200	18,400	11.10	25.83	17.46
2300	19,520	11.17	26.32	17.84
2400	20,640	11.23	26.80	18.20
2500	21,760	11.29	27.26	18.56
2600	22,890	11.33	27.70	18.90
2700	24,030	11.36	28.13	19.23
2800	25,170	11.38	28.55	19.56
2900	26,300	11.39	28.95	19.88
3000	27,440	11.39	29.33	20.18

[a] $H°_T - H°_{298.15} = -2.1417 \times 10^3 + 6.7057T + 1.5942 \times 10^{-3}T^2 - 1.8076 \times 10^{-7}T^3 + 1.6148 \times 10^3/T$ (298–1800°K, cal/mole, $\pm 0.4\%$) ; mol. wt. $= 98.911 : H°_{298.15} - H°_0$ $= 1270$ cal/mole. A randomization entropy of 1.37 eu is not included.
[b] Data evaluated by Storms (3).

TABLE XI

THERMAL FUNCTIONS OF $NbC_{0.75}$ [a, b]

T (°K)	$H_T^\circ - H_{298}^\circ$ (cal/mole)	C_p° (cal/mole-deg)	S_T° (cal/mole-deg)	$-(F_T^\circ - H_{298}^\circ)/T$ (cal/mole-deg)
298.15	0.00	8.100	7.720	7.720
300	15.01	8.123	7.770	7.720
400	876.7	9.018	10.24	8.051
500	1807	9.558	12.32	8.703
600	2784	9.958	14.10	9.457
700	3797	10.29	15.66	10.23
800	4841	10.59	17.05	11.00
900	5913	10.86	18.31	11.74
1000	7012	11.12	19.47	12.46
1100	8136	11.36	20.54	13.15
1200	9284	11.60	21.54	13.80
1300	10,460	11.84	22.48	14.44
1400	11,650	12.07	23.37	15.04
1500	12,870	12.29	24.21	15.63
1600	14,110	12.51	25.01	16.19
1700	15,370	12.72	25.77	16.73
1800	16,650	12.94	26.50	17.25
1900	17,960	13.15	27.21	17.76
2000	19,280	13.35	27.89	18.25
2100	20,630	13.56	28.55	18.72
2200	21,990	13.76	29.18	19.18
2300	23,380	13.96	29.80	19.63
2400	24,790	14.15	30.39	20.07
2500	26,210	14.35	30.98	20.49
2600	27,660	14.54	31.54	20.91
2700	29,120	14.73	32.10	21.31
2800	30,600	14.91	32.63	21.71
2900	32,100	15.10	33.16	22.09
3000	33,620	15.28	33.68	22.47

[a] $H_T^\circ - H_{297.15}^\circ = -3.2096 \times 10^3 + 8.8844T + 1.2409 \times 10^{-3}T^2 - 3.8370 \times 10^{-8}T^3 + 1.3460 \times 10^5/T$ (298–1800°K, cal/mole, ±0.2%); mol. wt. = 101.914; $H_{298.15}^\circ - H_0^\circ$ = 1305 cal/mole. A randomization entropy of 1.12 eu is not included.

[b] Data evaluated by Storms (3).

TABLE XII

THERMAL FUNCTIONS OF $NbC_{0.87}$ [a, b]

T (°K)	$H_T^\circ - H_{298}^\circ$ (cal/mole)	C_p° (cal/mole-deg)	S_T° (cal/mole-deg)	$-(F_T^\circ - H_{298}^\circ)/T$ (cal/mole-deg)
298.15	0.00	8.510	7.980	7.980
300	15.77	8.537	8.033	7.980
400	925.2	9.536	10.64	8.329
500	1909	10.10	12.84	9.017
600	2940	10.50	14.71	9.814
700	4006	10.81	16.36	10.63
800	5101	11.08	17.82	11.44
900	6222	11.33	19.14	12.23
1000	7366	11.57	20.34	12.98
1100	8534	11.79	21.46	13.70
1200	9724	12.01	22.49	14.39
1300	10,940	12.22	23.46	15.05
1400	12,170	12.44	24.38	15.68
1500	13,420	12.65	25.24	16.29
1600	14,700	12.86	26.06	16.88
1700	15,990	13.06	26.85	17.44
1800	17,310	13.27	27.60	17.98
1900	18,650	13.48	28.32	18.51
2000	20,010	13.69	29.02	19.02
2100	21,390	13.90	29.69	19.51
2200	22,790	14.11	30.35	19.99
2300	24,210	14.32	30.98	20.45
2400	25,650	14.53	31.59	20.90
2500	27,110	14.74	32.19	21.34
2600	28,600	14.96	32.77	21.77
2700	30,110	15.17	33.34	22.19
2800	31,630	15.38	33.90	22.60
2900	33,180	15.60	34.44	23.00
3000	34,750	15.82	34.97	23.39

[a] $H_T^\circ - H_{298.15}^\circ = -3.5738 \times 10^3 + 9.8318T + 9.2474 \times 10^{-4}T^2 + 1.6788 \times 10^{-8}T^3 + 1.6691 \times 10^5/T$ (298–1800°K, cal/mole, $\pm 0.2\%$) ; mol. wt. $= 103.355$; $H_{298.15}^\circ - H_0^\circ = 1350$ cal/mole. A randomization entropy of 0.77 eu is not included.

[b] Data evaluated by Storms (3).

TABLE XIII

THERMAL FUNCTIONS OF $NbC_{0.98}$ [a, b]

T (°K)	$H_T^\circ - H_{298}^\circ$ (cal/mole)	C_p° (cal/mole-deg)	S_T° (cal/mole-deg)	$-(F_T^\circ - H_{298}^\circ)/T$ (cal/mole-deg)
298.15	0.00	8.790	8.290	8.290
300	16.29	8.825	8.344	8.290
400	971.6	10.12	11.08	8.655
500	2021	10.81	13.42	9.381
600	3126	11.26	15.44	10.23
700	4269	11.59	17.20	11.10
800	5442	11.86	18.76	11.96
900	6640	12.10	20.18	12.80
1000	7861	12.30	21.46	13.60
1100	9101	12.49	22.64	14.37
1200	10,360	12.67	23.74	15.11
1300	11,630	12.84	24.76	15.81
1400	12,930	13.00	25.72	16.48
1500	14,230	13.15	26.62	17.13
1600	15,560	13.30	27.47	17.75
1700	16,890	13.44	28.28	18.34
1800	18,250	13.58	29.05	18.92
1900	19,610	13.71	29.79	19.47
2000	20,990	13.84	30.50	20.00
2100	22,380	13.97	31.38	20.52
2200	23,780	14.09	31.83	21.02
2300	25,200	14.21	32.46	21.50
2400	26,620	14.32	33.07	21.97
2500	28,061	14.43	33.65	22.43
2600	29,510	14.54	34.22	22.87
2700	30,968	14.64	34.77	23.30
2800	32,437	14.74	35.31	23.72
2900	33,916	14.84	35.82	24.13
3000	35,404	14.93	36.33	24.53

[a] $H_T^\circ - H_{298.15}^\circ = -4.0918 \times 10^3 + 10.8561T + 9.1724 \times 10^{-4}T^2 - 5.2003 \times 10^{-8}T^3 + 2.3105 \times 10^5/T$ (298–3000°K, cal/mole, ±0.3%); mol. wt. = 104.676; $H_{289.15}^\circ - H_0^\circ$ = 1390 cal/mole.

[b] Data evaluated by Storms (3).

TABLE XIV

THERMAL FUNCTIONS OF $TaC_{0.99}$ [a, b]

T (°K)	$H_T^\circ - H_{298}^\circ$ (cal/mole)	C_p° (cal/mole-deg)	S_T° (cal/mole-deg)	$-(F_T^\circ - H_{298}^\circ)/T$ (cal/mole-deg)
298.15	0.00	8.764	10.10	10.10
300	16.24	8.794	10.15	10.10
400	958.9	9.931	12.86	10.46
500	1987	10.58	15.15	11.18
600	3068	11.03	17.12	12.01
700	4190	11.39	18.85	12.86
800	5345	11.69	20.39	13.71
900	6528	11.97	21.78	14.53
1000	7738	12.22	23.06	15.32
1100	8971	12.45	24.23	16.08
1200	10,228	12.68	25.33	16.80
1300	11,506	12.89	26.35	17.50
1400	12,805	13.09	27.31	18.16
1500	14,124	13.28	28.22	18.81
1600	15,462	13.47	29.08	19.42
1700	16,818	13.65	29.91	20.01
1800	18,193	13.83	30.69	20.59
1900	19,584	14.00	31.44	21.14
2000	20,992	14.16	32.17	21.67
2100	22,417	14.32	32.86	22.19
2200	23,857	14.47	33.53	22.69
2300	25,312	14.62	34.18	23.17
2400	26,781	14.76	34.80	23.65
2500	28,264	14.90	35.41	24.10
2600	29,761	15.03	36.00	24.55
2700	31,271	15.16	36.57	24.98
2800	32,793	15.28	37.12	25.41
2900	34,327	15.40	37.66	25.82
3000	35,872	15.51	38.18	26.22

[a] $H_T - H_{298.15}^\circ = -3.7468 \times 10^3 + 10.1132T - 1.2668 \times 10^{-3}T^2 - 8.0868 \times 10^{-8}T^3 + 1.8517 \times 10^4/T$ (298–3000°K, cal/mole) ; mol. wt. = 192.84.

[b] Data evaluated by Storms (3).

V. Theoretical Treatments

At very high temperatures, the experimental determinations of thermo-dynamic data become increasingly more difficult and costly, not only because of the experimental difficulties associated with the high temperatures, but also because of the difficulties in controlling the composition of samples which do not vaporize congruently. Because of these experimental problems and the need for partial free energies as a function of composition, a number of estimation techniques have been developed. Storms (3) suggests that the thermodynamic values of group V carbides can be estimated as a function of composition from the experimentally determined data in the NbC_{1-x} system. For the other carbides and for the nitrides, however, this technique cannot be used.

Kaufman *et al.* (36–38) and Chang (2) have extended the Schottky–Wagner theory for nonstoichiometric alloys to carbides for the following purposes:

(1) Estimation of the partial molar free energies $\bar{G}_{Me}$ and $\bar{G}_C$ or $\bar{G}_N$ as a function of composition.

(2) Estimation of metal and nonmetal vapor pressures as a function of composition and temperature.

(3) Estimation of compositions for congruent vaporization as a function of temperature.

(4) Estimation of binary phase diagrams.

(5) Estimation of ternary and higher order phase diagrams.

Use of this method requires several drastic assumptions about random mixing of vacancies and an energy of vacancy formation independent of composition.

The method can be readily adapted for estimating vapor pressures of the metal and nonmetal as a function of composition and pressure. The equations for calculating the pressure as a function of composition for $Me_{1-x}C_x$ are divided into two groups:

at the stoichiometric composition

$$RT \ln \{P_{Me}^\sigma[0.5]/P_{Me}^\circ{}^\circ\} = G_{Me^+} - RT \ln 2\alpha, \qquad (11)$$

$$RT \ln \{P_C^\sigma[0.5]/P_C^\circ\} = G_{C^+} - RT \ln 2\alpha, \qquad (12)$$

and at composition $x < 0.5$

$$RT \ln \left\{\frac{P_{Me}^\sigma[x]}{P_{Me}^\circ}\right\} = -G_{Me^+} + RT \ln \left[\frac{1-2x}{4(1-x)\alpha^2}\right], \qquad (13)$$

$$RT \ln \left\{\frac{P_C^\sigma[x]}{P_C^\circ}\right\} = -G_{C^+} + RT \ln \left[\frac{x}{1-2x}\right]. \qquad (14)$$

TABLE XV

THERMAL FUNCTIONS OF $CrC_{2/3}$ [a, b]

T (°K)	$H_T^\circ - H_{298}^\circ$ (cal/mole)	C_p° (cal/mole-deg)	S_T° (cal/mole-deg)	$-(F_T^\circ - H_{298}^\circ)/T$ (cal/mole-deg)
298.15	0.00	7.84	6.81	6.81
300	14.53	7.87	6.86	6.81
400	870.7	9.11	9.31	7.14
500	1819	9.81	11.43	7.79
600	2826	10.30	13.26	8.55
700	3876	10.70	14.88	9.34
800	4963	11.04	16.33	10.13
900	6083	11.35	17.65	10.89
1000	7234	11.65	18.86	11.63
1100	8413	11.94	19.99	12.34
1200	9621	12.22	21.04	13.02
1300	10,860	12.49	22.02	13.67
1400	12,120	12.76	22.96	14.30
1500	13,410	13.04	23.85	14.91
1600	14,730	13.30	24.70	15.50
1700	16,070	13.57	25.51	16.06
1800	17,440	13.84	26.30	16.61
1900	18,840	14.12	27.05	17.14
2000	20,260	14.39	27.78	17.65
2100	21,720	14.66	28.49	18.15
2200	23,200	14.93	29.18	18.64
2300	24,700	15.21	29.85	19.11
2400	26,240	15.48	30.50	19.57
2500	27,800	15.76	31.14	20.02

[a] $H_T^\circ - H_{298.15}^\circ = -3.6074 \times 10^3 + 9.4443T + 1.1635 \times 10^{-3}T^2 + 2.8241 \times 10^{-8}T^3 + 2.0496 \times 10^5/T$ (298–1600°K, cal/mole, ±0.7%) ; mol. wt. = 60.00.

[b] Data evaluated by Storms (3).

In Eqs. (11)–(14), P_{Me}° and P_C° are the vapor pressures of pure carbon and pure metal at temperature T, P_{Me}^σ and P_C^σ are the pressures in the monocarbide phase σ, G_{Me^+} and G_{C^+} are the temperature-dependent free energies of formation of metal-atom vacancies and carbon-atom vacancies, respectively. The parameter α is the ratio of vacant metal sites to total lattice sites at the stoichiometric composition ($x = 0.5$). The free energies to form vacancies and α are related to the enthalpy of formation at 0°K of the

TABLE XVI

THERMAL FUNCTIONS OF CrC$_{6/23}$ [a, b]

T (°K)	$H_T^\circ - H_{298}^\circ$ (cal/mole)	C_p° (cal/mole-deg)	S_T° (cal/mole-deg)	$-(F_T^\circ - H_{298}^\circ)/T$ (cal/mole-deg)
298.15	0.00	6.487	6.325	6.325
300	12.02	6.510	6.365	6.325
400	711.0	7.358	8.370	6.593
500	1471	7.797	10.06	7.122
600	2266	8.091	11.51	7.736
700	3087	8.327	12.78	8.368
800	3931	8.545	13.90	8.991
900	4796	8.761	14.92	9.594
1000	5683	8.986	15.86	10.17
1100	6593	9.224	16.73	10.73
1200	7529	9.480	17.54	11.27
1300	8490	9.754	18.31	11.78
1400	9480	10.05	19.04	12.27
1500	10,500	10.37	19.75	12.75
1600	11,550	10.70	20.43	13.20
1700	12,640	11.06	21.08	13.65
1800	13,770	11.45	21.73	14.08
1900	14,930	11.85	22.36	14.50
2000	16,140	12.28	22.98	14.91

[a] $H_T^\circ - H_{298.15}^\circ = -3.0412 \times 10^3 + 8.3629T - 2.0104 \times 10^{-4}T^2 + 3.9683 \times 10^{-7}T^3 + 1.6551 \times 10^5/T$ (298–1700°K, cal/mole, ±0.3%); mol. wt. = 55.13.
[b] Data evaluated by Storms (3).

stoichiometric phase and the free energy of formation of the stoichiometric phase at the temperature of interest by the relationships:

$$RT \ln \alpha = 2\Delta H_f^\sigma[0°K] \approx 2\Delta H_f^\sigma[298°K], \qquad (15)$$

and

$$2\Delta G^\sigma[x = 0.5] = -G_{Me^+} - G_{C^+} - 2\,RT \ln 2\alpha. \qquad (16)$$

Equation (16) is the conditional equation in the Schottky–Wagner theory which minimizes the free energy at any given pressure, temperature and composition. If we furthermore assume that the boundary of the monocarbides or mononitrides in equilibrium with graphite or nitrogen occurs at $x = 0.5$, the chemical potential of carbon in the monocarbide equals that of graphite and then we can evaluate G_{C^+} and G_{Me^+} by

TABLE XVII

THERMAL FUNCTIONS OF $CrC_{3/7}$ [a, b]

T (°K)	$H_T^\circ - H_{298}^\circ$ (cal/mole)	C_p° (cal/mole-deg)	S_T° (cal/mole-deg)	$-(F_T^\circ - H_{298}^\circ)/T$ (cal/mole-deg)
298.15	0.00	7.131	6.857	6.857
300	13.22	7.160	6.901	6.857
400	786.7	8.169	9.120	7.153
500	1631	8.661	11.00	7.739
600	2513	8.973	12.61	8.420
700	3423	9.222	14.01	9.121
800	4357	9.454	15.26	9.811
900	5314	9.695	16.39	10.48
1000	6297	9.954	17.42	11.12
1100	7306	10.24	18.38	11.74
1200	8346	10.56	19.29	12.33
1300	9419	10.91	20.14	12.90
1400	10,530	11.29	20.97	13.45
1500	11,680	11.71	21.76	13.97
1600	12,870	12.17	22.53	14.49
1700	14,110	12.67	23.28	14.98
1800	15,410	13.20	24.02	15.46
1900	16,750	13.77	24.75	15.93
2000	18,160	14.38	25.47	16.39

[a] $H_T^\circ - H_{298.15}^\circ = -3.5948 \times 10^3 + 9.8464T - 8.2590 \times 10^{-4}T^2 + 6.5761 \times 10^{-7}T^3 + 2.1319 \times 10^5/T$ (298–1600°K, cal/mole, $\pm 0.3\%$) ; mol. wt. = 57.14.
[b] Data evaluated by Storms (3).

$$G_{C^+} = -RT \ln 2\alpha, \tag{17}$$

and

$$G_{Me^+} = 2\Delta G_{f, 0.5} - RT \ln 2 - 2\Delta H_{f, 0.5}[0°K]. \tag{18}$$

Thus if $\Delta H_{f,0.5}[0°K]$ and $\Delta G_{f,0.5}(T)$ are known, then α, G_{Me^+}, and G_{C^+} are determined by Eqs. (16)–(18) and these latter parameters then define the partial and integral free energies as a function of composition.

In Figs. 8 and 9, we show computer calculations of the vapor pressures of several carbides as a function of composition at 3000°K. The only experimental input into the computer program is the free energy of formation at stiochiometry of the monocarbide, the vapor pressure of graphite and the pure metal element at the specified temperature, and the heat of formation of the stoichiometric monocarbide at 298.15°K.

Figure 10 compares the calculated vapor pressures of TiC and ZrC with experimental data (3). For ZrC the agreement is fair but for TiC the agreement

TABLE XVIII

HEAT CONTENT OF $MoC_{1/2}$ [a, b]

T (°K)	$H_T^\circ - H_{298}^\circ$ (cal/mole)	C_p° (cal/mole-deg)	S_T° (cal/mole-deg)	$-(F_T^\circ - H_{298}^\circ)/T$ (cal/mole-deg)
298.15	0.00	7.190	7.865	7.86
300	13.32	7.212	7.91	7.86
400	781.0	8.049	10.11	8.16
500	1612	8.531	11.96	8.74
600	2483	8.871	13.55	9.41
700	3384	9.140	14.94	10.10
800	4309	9.368	16.17	10.79
900	5256	9.571	17.29	11.45
1000	7207	9.925	19.25	12.69
1200	8207	10.08	20.12	13.28
1300	9223	10.23	20.93	13.83
1400	10,253	10.37	21.69	14.37
1500	11,297	10.50	22.41	14.88
1600	12,354	10.63	23.09	15.37
1700	13,422	10.74	23.74	15.85
1800	14,502	10.85	24.36	16.30
1900	15,593	10.96	24.95	16.74
2000	16,694	11.05	25.51	17.17
2100	17,803	11.14	26.06	17.58
2200	18,922	11.23	26.58	17.97
2300	20,048	11.31	27.08	18.36
2400	21,183	11.37	27.56	18.73
2500	22,324	11.44	28.02	19.10
2600	23,471	11.50	28.47	19.45
2700	24,624	11.55	28.91	19.79
2800	25,782	11.60	29.33	20.12
2900	26,944	11.64	29.74	20.45
3000	28,110	11.69	30.13	20.76

[a] $H_T^\circ - H_{298.15}^\circ = -2.9502 \times 10^3 + 8.0920T + 1.0471 \times 10^{-3}T^2 - 9.9403 \times 10^{-8}T^3 + 1.3333 \times 10^5/T$.

[b] Data evaluated by Storms (32).

is poor. In both systems the vapor pressure of the metal atoms is over-estimated in the calculation. One reason for the discrepancy is clearly the assumption that the free energies G_{Me^+} and G_{C^+} to form metal and carbon vacancies are independent of composition. If the Debye temperature is used as an indication of bond strength then that strength becomes greater as

TABLE XIX

HEAT CONTENT OF WC [a, b]

T (°K)	$H_T^\circ = H_{298}^\circ$ (cal/mole)	C_p° (cal/mole-deg)	$S_T^\circ - S_{298.15}^\circ$ (cal/mole-deg)
298.15	0.00	9.500	0.000
300	17.58	9.509	0.059
400	990.6	9.938	2.855
500	2004	10.31	5.11
600	3053	10.66	7.02
700	4135	10.99	8.69
800	5250	11.29	10.18
900	6394	11.58	11.53
1000	7566	11.86	12.76
1100	8765	12.12	13.90
1200	9989	12.36	14.97
1300	11,237	12.59	15.97
1400	12,506	12.80	16.91
1500	13,797	13.00	17.80
1600	15,106	13.18	18.64
1700	16,432	13.35	19.45
1800	17,775	13.50	20.22
1900	19,132	13.64	20.95
2000	20,503	13.76	21.65
2100	21,885	13.87	22.33
2200	23,277	13.97	22.97
2300	24,677	14.04	23.60
2400	26,085	14.11	24.20
2500	27,499	14.16	24.77
2600	28,916	14.19	25.33
2700	30,337	14.21	25.87
2800	31,758	14.22	26.38
2900	33,180	14.21	26.88
3000	34,599	14.18	27.36

[a] $H_T^\circ - H_{298.15}^\circ = -2.7595 \times 10^3 + 8.5025T + 2.0550 \times 10^{-3}T^2 - 2.4625 \times 10^{-7}T^3 + 1.4424 \times 10^4/T$.

[b] Data evaluated by Storms (32).

stoichiometry is approached. This correction to G_{Me^+} would tend to increase the discrepancy between calculation and experiment since the calculated values of P_{Me^+} would increase still further. One must, therefore, consider the compositional dependence of the entropy term in forming vacancies.

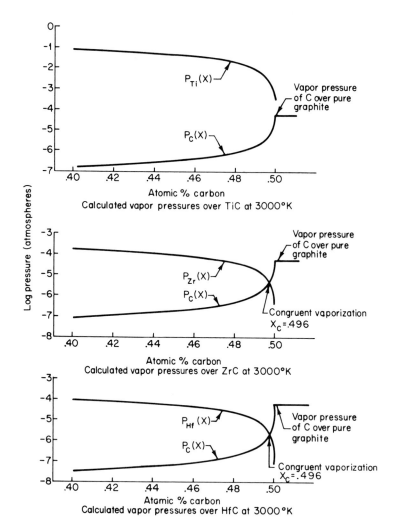

Fig. 8. Calculated vapor pressures over TiC_{1-x}, ZrC_{1-x}, and HfC_{1-x} at 3000°K as a function of composition.

This term apparently increases with stoichiometry deviations at a rate sufficient enough to offset the enthalpy term. The net result is that G_{Me^+} actually increases with deviations from stoichiometry. The situation will become clearer as more experimental data on vapor pressures are reported and as more measurements of the enthalpy of vacancy formation are determined. An excellent review of current vapor pressure measurements has been given by Storms (39).

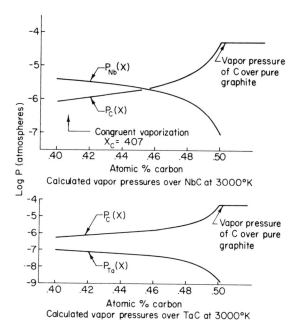

Fig. 9. Calculated vapor pressures over NbC_{1-x} and TaC_{1-x} at 3000°K as a function of composition.

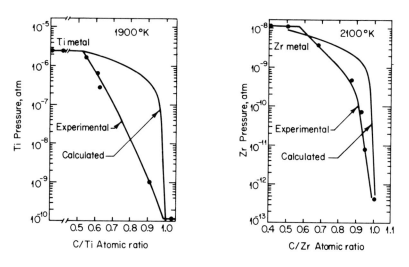

Fig. 10. Comparison of calculated and observed vapor pressures over TiC_{1-x} and ZrC_{1-x}.

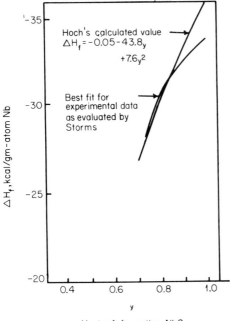

Fig. 11. Comparison of theoretical values for the heat of formationa as a function of composition with experimental values. [The calculated values are from Hoch (41).]

Heat of formation NbC$_y$

   Hoch (40, 41) has also applied regular solution theory to the treatment of interstitial solid solutions. He groups together first, second, and third nearest-neighbor bonds into one interaction energy E_{ij}, which is assumed to be independent of composition. His method is similar to that of Kaufman et al. and Chang. Introducing the E_{ij}'s into a grand partition function, Hoch then develops expressions for the activities of each chemical species in the solution and uses experimental activity measurements to evaluate the E_{ij}'s. Once the E_{ij}'s are known, other parameters such as the heats of formation, can be calculated with fair success. Figure 11 compares the heats of formation for NbC$_{1-x}$ calculated by Hoch with the experimental values. Since the method is semiempirical in the sense that some experimental input has gone into evaluating the E_{ij}'s, a high degree of success should be required of the theory and as the figure shows the agreement is good over the carbon-poor side of the NbC$_{1-x}$ phase field but only fair at the stoichiometric composition.

   Some degree of success has been achieved with the regular solution model for predicting phase diagrams (37) and for using the experimentally determined phase diagrams to predict partial molar free energies (2). Since most phase diagrams are now known, however, the former approach of predicting them no longer seems necessary. The latter approach of using the phase diagrams to predict partial molar quantities has not been adequately tested with experimental data.

We have already noted the difficulties of finding suitable theoretical estimations of standard state entropies and enthalpies at 298°K. This fact, together with the inadequacies of the Schottky–Wagner model for estimating vapor pressures, demonstrates the need for better theoretical treatments of the thermodynamics of these phases. Until there is either better theoretical predications or more experimental information, particularly on non-stoichiometric phases, the thermodynamics of carbides and nitrides will remain incompletely known.

References

1. H. L. Schick, ed., "Thermodynamics of Certain Refractory Compounds," Vols. 1 and 2. Academic Press, New York, 1966.
2. Y. A. Chang, "Thermodynamic Properties of Group IV, V, and VI Binary Transition Metal Carbides," AFML-TR-65-2, Part IV, Vol. I (1965).
3. E. K. Storms, "The Refractory Carbides." Academic Press, New York, 1967.
4. G. V. Samsonov, "High-Temperature Materials, Properties Index." Plenum Press, New York, 1964.
5. K. K. Kelley, "Entropies of the Elements and Inorganic Compounds." U.S. Printing Office, Washington, D.C., 1961; see also "High Temperature Heat-Capacity and Entropy Data for the Elements and Inorganic Compounds." U.S. Printing Office, Washington, D.C., 1960.
6. K. K. Kelley, in "Selected Values of Thermodynamic Properties of Metals and Alloys" (R. Hultgren, ed.), Supplement. Wiley, New York, 1963.
7. D. R. Stull, dir., "JANAF Thermochemical Tables." Dow Chem. Co., Midland, Michigan, 1965.
8. N. Pessall, J. K. Hulm, and M. S. Walker, Final Rep., Westinghouse Research Laboratories, AF 33(615)-2729 (1967).
9. L. E. Toth, M. Ishikawa, and Y. A. Chang, Acta Met. 16, 1183 (1968).
10. L. E. Toth, J. Zbasnik, Y. Sato, and W. Gardner, in "Anistropy in Single-Crystal Refractory Compounds" (F. W. Vahldiek and S. A. Mersol, eds.), Vol. I, p. 249. Plenum Press, New York, 1968.
11. L. E. Toth and J. Zbasnik, Acta Met. 16, 1177 (1968).
12. Y. A. Chang, L. E. Toth, and Y. S. Tyan, to be published, Trans. AIME.
13. W. Nernst and F. A. Lindemann, Z. Elektrochem. 17, 817 (1911).
14. L. Kaufman, Trans. AIME 224, 1006 (1964).
15. Y. A. Chang, Trans. AIME 239, 1685 (1967).
16. A. D. Mah, U.S., Bur. Mines, Rep. Investi. RI-6518 (1964).
17. E. J. Huber, Jr., E. L. Head, C. E. Holley, Jr., E. K. Storms, and N. H. Krikorian, J. Phys. Chem. 65, 1846 (1961).
18. A. D. Mah and B. J. Boyle, J. Amer. Chem. Soc. 77, 6512 (1955).
19. A. N. Kornilov, V. Ya. Leonidov, and S. M. Skuratov, Vestn. Mosc. Univ., Khim. 17, No. 6, 48 (1962).
20. F. G. Kusenko and P. V. Gel'd, Izv. Sib. Otd. Akad. Nauk SSSR No. 2, 46 (1960).
21. A. N. Kornilov, I. D. Zaikin, S. M. Skuratov, and G. P. Shveikin, Zh. Fiz. Khim. 40, 1070 (1966).

22. W. L. Worrell and J. Chipman, *J. Phys. Chem.* **68**, 860 (1964).
23. E. J. Huber, Jr., E. L. Head, C. E. Holley, Jr., and A. L. Bowman, *J. Phys. Chem.* **67**, 793 (1963).
24. A. N. Kornilov, I. D. Zaikin, S. M. Skuratov, L. B. Dubrovskaya, and G. P. Shveikin, *Zh. Fiz. Khim.* **38**, 702 (1964).
25. P. M. McKenna, *Ind. Eng. Chem. Soc.* **28**, 767 (1936).
26. G. L. Humphrey, *J. Amer. Chem. Soc.* **76**, 978 (1954).
27. A. D. Mah, *U.S. Bur. Mines, Rep. Invest.* **RI-6663** (1965).
28. V. I. Smirnova and B. F. Ormont, *Zh. Fiz. Khim.* **30**, 1327 (1956).
29. V. I. Zhelankin and V. S. Kutsev, *Zh. Fiz. Khim.* **38**, 562 (1964).
30. E. K. Storms, LAMS-2674, Part II (1962).
31. P. Costa and R. R. Conte, *in* "Compounds of Interest in Nuclear Reactor Technology" (J. T. Waber, P. Chiotti, and W. N. Miner, eds.), Inst. Metals Div., Spec. Rep. No. 13, p. 3. Edwards, Ann Arbor, Michigan, 1964.
32. E. K. Storms, private communication (1969). The author is indebted to Ed Storms for allowing publication of these tables prior to his publishing them elsewhere.
33. C. G. Maier and K. K. Kelley, *J. Amer. Chem. Soc.* **54**, 3243 (1932).
34. C. H. Shomate, *J. Amer. Chem. Soc.* **66**, 928 (1944).
35. T. G. Godfrey and J. M. Leitnaker, ORNL-TM-1599 (1966).
36. L. Kaufman, H. Bernstein, and A. Sarney, ASD-TR-6-445, Part IV (1963).
37. L. Kaufman and A. Sarney, *in* "Compounds of Interest in Nuclear Reactor Technology" (J. T. Waber, P. Chiotti, and W. N. Miner, eds.), Inst. Metals Div., Spec. Rep. No. 13, p. 267. Edwards, Ann Arbor, Michigan, 1964.
38. L. Kaufman and E. V. Clougherty, *in* "Metallurgy at High Temperature and High Pressures" (K. A. Gschneider Jr., M. T. Hepworth, and N. A. D. Parlee, eds.), *AIME Symp.*, p. 322, 1963.
39. E. K. Storms *in* "Fundamentals of Refractory Compounds" (H. H. Hausner and M. G. Bowman, eds.), p. 67. Plenum Press, New York, 1968.
40. M. Hoch, *Trans. AIME* **230**, 138 (1964).
41. M. Hoch, in "Anistropy in Single-Crystal Refractory Compounds" (F. W. Vahldiek and S. A. Mersol, eds.), Vol. I, p. 163. Plenum Press, New York, 1968.

5
Mechanical Properties

I. Introduction

Transition metal carbides are potentially important materials in engineering applications because of their great strength and hardness. In commercial applications they are used extensively, for example, as cutting tool bits, wear-resistant surface finishes, dies, and automobile tire studs. They may be either dispersed in a matrix material to increase the strength of the matrix or utilized alone because of their intrinsic strength. The usefulness of these materials is limited by their brittleness at ordinary temperatures. However, at high temperatures they become quite ductile yet retain much of their strength when properly alloyed.

The need for improved materials for high-temperature structural applications has stimulated much of the research on the mechanical properties of carbides. The carbides are of interest in these applications for a number of reasons including their very high melting points, great strength, and ability to deform plastically on slip systems analogous to fcc metals at high temperatures. The early research and development of these materials centered on polycrystalline carbides and, in general, the results of this research were disappointing because the materials used proved to be very brittle. In retrospect, one can ascribe these poor properties to the use of highly porous materials; the pores acted as stress concentrators and initiation sites for fracture. The early research on sintered compacts was mainly confined to

141

determinations of Young's moduli, bend strengths, and microhardness. Much of this work has been reviewed by Westbrook and Stover (*1*). As these authors correctly point out, the available data on polycrystalline carbides is of limited value because it cannot be used to assess the intrinsic strengths of the carbides or the relative strengths of different carbides. The variable and limited nature of this data also raises questions about its reliability (*2*).

In the past five to ten years the emphasis on the mechanical properties of carbides has changed from the testing of sintered samples to the testing of fully dense single crystals or polycrystalline carbides obtained from the melt. Testing of these fully dense materials has resulted in more accurate values of the elastic constants and yield strengths. The experiments utilizing single crystals enabled plastic deformation mechanisms to be deduced. In this chapter we review the mechanical behavior studies conducted mainly on single crystals and where appropriate we include the results on polycrystalline materials. The nitrides are not included in this discussion because very little research has been conducted on their mechanical behavior.

The knowledge gained from these studies on the mechanical properties of single crystals of carbides can be used to develop better structural materials in engineering applications. Single crystals themselves are perhaps too difficult to prepare in large enough sizes to be of importance in commercial applications. Nevertheless, the research has shown that carbides plastically deform on slip systems analogous to fcc metals. The carbides therefore have a sufficient number of independent slip systems so that polycrystalline carbides can be made ductile. In fact, fully dense polycrystalline TiC exhibited 30% ductility at 1500°C and a yield strength which is 10 times greater than that of single crystals of the same composition and temperature (*3*). Utilization of fully dense fine grained TiC has also resulted in an improvement of the elastic moduli. Therefore, these materials in the form of fully dense fine-grained polycrystalline bodies hold considerable promise as high-temperature structural materials.

In discussing the mechanical behavior of carbides we have subdivided the subject into the topics—elastic deformation, plastic deformation, fracture, strengthening mechanisms, and hardness. The elastic behavior of solids is controlled primarily by the strength of the atomic bonds. If the type of atomic bonding in a particular solid is known we can predict some aspects of its elastic behavior, for example, the elastic moduli, and conversely, we can use information about elastic properties to help understand types of atomic bonding. Elastic moduli can also be used to predict theoretical strengths in the absence of imperfections and flaws. The higher the modulus the greater the theoretical strength. Often, however, the useful strength of a solid is limited by the motion of dislocations (plastic deformation) or by failure due to fracture. Fracture is usually controlled by small cracks, either internal or surface or by insufficient dislocation mobility to stop crack propagation. Thus a material may have a very high elastic modulus but a relatively low strength because it flows plastically at stress levels thousands of times smaller than the theoretical limit, or another material with a high modulus may be weak because it contains small surface or internal flaws which initate cracks and cause failure.

II. Elastic Behavior

To describe the elastic behavior of a solid three constants are involved: the bulk modulus K, the shear modulus G and Poisson's ratio v. Young's modulus E is also used frequently, although it is not a fundamental elastic parameter. For the carbides, as well as other ceramic materials, these parameters will be a function of the temperature, porosity, distribution of that porosity, composition, and vacancy concentration.

Not all of these parameters have been studied for the carbides, but there is sufficient evidence that the moduli are sensitive functions of both temperature and porosity. The temperature dependence of E, G, and v for a hot-pressed sample of $NbC_{0.97}$ of 94% theoretical density is shown in Fig. 1 (4). In the temperature region between 0 and 1600°C both E and G decrease by about 25% while v is relatively temperature insensitive. Both E and G follow the approximate temperature relationship (4)

$$E = E_0 - BT \exp(-T_0/T) \tag{1}$$

and

$$G = G_0 - CT \exp(-T_0'/T) \tag{2}$$

where B, C, T_0, T_0' are constants characteristic of the material. These relationships are the same as followed by several oxides. For the oxides these relationships have received theoretical justification by Anderson (5): T_0 is related to the Debye temperature and B is related to the Grüneisen constant. These results are in substantial agreement with those of Brenton et al. (6) for $NbC_{0.97}$ with 96% theoretical density. For G they find a linear dependence upon T.

The moduli are also sensitive functions of the porosity. Speck and Miccioli (4) have investigated the effect of porosity on E and G for hot-pressed NbC and find the following relationships

$$E = 74.4 \, (1 - 1.9P) \times 10^6 \text{ psi} \tag{3}$$

and

$$G = 30.6 \, (1 - 0.78P) \times 10^6 \text{ psi} \tag{4}$$

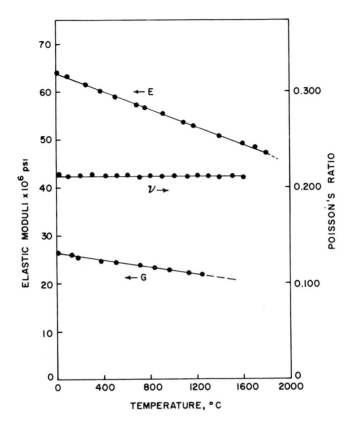

Fig. 1. The temperature dependence of E, G, and ν for $NbC_{0.97}$ 94% theoretical density. [After Speck and Miccioli (4).]

where P is the porosity. These results are shown in Fig. 2. The effects of the shape and distribution of that porosity were not determined.

In view of the sensitivity of E and G to the porosity and the difficulty of estimating corrections for that porosity, it is little wonder that the earlier literature (1, 2, 7) on E and G values for carbides contains a large number of widely differing values. Most of these earlier values were obtained on highly porous samples and the means for correcting for that porosity were unknown. To help eliminate problems associated with porosity, elastic constants have been measured recently on single crystals and nearly 100% dense hot-pressed samples. Much of this information has been derived by pulse–echo techniques.

Gilman and Roberts (8), who first measured the elastic constants of single

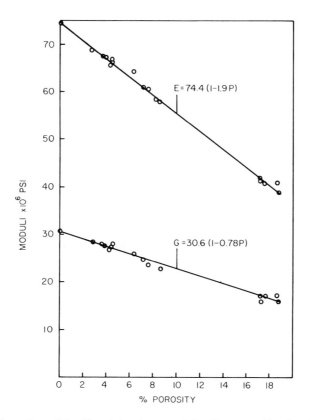

Fig. 2. Young's modulus E and the shear modulus G are sensitive functions of the porosity in sintered samples of $NbC_{0.97}$. This behavior is fairly typical of ceramic materials. [After Speck and Miccioli (*4*).]

crystals of TiC (unspecified composition), found that TiC is nearly isotropic; the anisotropy parameter $2C_{44}/(C_{11} - C_{12})$ equals 0.91. Values of the elastic constants are listed in Table I (*8–12*). Chang and Graham (*9*) determined the elastic constants of single crystals of $ZrC_{0.94-0.89}$ and $TiC_{0.91}$ and they found that the elastic constants C_{11}, C_{12}, and C_{44} vary by only a few percent over the temperature interval 4–298°K.

Even with these measurements on single crystals, there are some serious questions about how accurately the elastic constants are known. For TiC the first three entries in Table I are in good agreement with each other but not with the fourth entry. Likewise the C_{12} values for ZrC_{1-x} are in poor agreement. Whether these discrepancies are due to the use of slightly different experimental techniques or due to sample variations is not clear because not

TABLE I

ELASTIC CONSTANTS OF SINGLE CRYSTALS OF TiC, ZrC, AND TaC AT 298°K [a]

Compound	C_{11}	C_{12}	C_{44}	Ref.
$TiC_{0.91}$	5.145 ± 0.0005	1.060 ± 0.002	1.788 ± 0.002	(a)
TiC	5.245	0.980	1.809	(b)
TiC	5.00	1.13	1.75	(c)
TiC	3.891	0.433	2.032	(d)
$ZrC_{0.94-0.89}$	4.720 ± 0.0005	0.987 ± 0.0005	1.593 ± 0.002	(a)
ZrC	4.280	0.408	1.464	(b)
$TaC_{0.90}$	5.05 ($\pm 2\%$)	0.73–0.91 ($\pm 25\%$)	0.79 ($\pm 8\%$)	(e)
$VC_{0.84}$	5.00	2.92	1.55	(f)

[a] All values have the units 10^{12} dyne/cm^2

References for Table I

a. R. Chang and L. J. Graham, *J Appl. Phys.* **37**, 3778 (1966).
b. B. T. Bernstein, unpublished data, reported by Chang and Graham (ref. a).
c. J. J. Gilman and B. W. Roberts, *J. Appl. Phys.* **32**, 1405 (1961).
d. J. deKlerk, *Rev. Sci. Instr.* **36**, 1540 (1965).
e. R. W. Bartlett and C. W. Smith, *J. Appl. Phys.* **38**, 5428 (1967).
f. J. Martin, RIAS, private communication to G. Hollox (1969).

all investigators characterized their samples. In Table II an additional comparison is made using instead the values of E, G, and v which can be calculated from the values of C_{11}, C_{12}, and C_{44}. Included in Table II are values of E, G, and v derived from pulse–echo measurements by Speck and Miccioli (*4*) and Brown *et al.* (*13*, *14*) on hot-pressed samples and also the values tabulated by Lynch *et al.* (*2*) on self-bonded sintered compacts. Generally, there are very large variations in the reported values of the elastic constants: TiC, $E = 39$–67×10^6 psi; HfC, $E = 46$–67×10^6 psi; TaC, $E = 44$–78×10^6 psi. Certain measurements do, however, appear more reliable than others. The measurements of Chang and Graham (*9*) on $TiC_{0.91}$ are in good agreement with several other entries in Table I and also their values for ZrC_{1-x} are in agreement with the measurement on hot-pressed ZrC of Brown and Kempter (*13*). Likewise the values of Brown and Kempter (*13*) and those of Speck and Miccioli (*4*) on $NbC_{0.97}$ are in good agreement. Considering the measurements of these three groups to be reliable we arrive at the following preferred values for $E(10^6$ psi): $TiC_{0.91}$, $E = 65$; $ZrC_{0.94}$, $E = 59$; $HfC_{0.97}$, $E = 67$; $NbC_{0.96}$, $E = 71$; TaC, $E = 78$, and WC, $E = 90$. We should note, however, the serious discrepancy between values for TaC and also take into account that several of the samples of Brown *et al.* (*14*) had large values for the porosity. Thus, in using these values, one must keep in mind that a few of the values in Table II are presently not accurately known.

One area that definitely needs further investigation on well-characterized samples is the effect of the carbon-to-metal ratio in the monocarbides on the elastic constants. Chang and Graham observed only a 1% variation in values between $ZrC_{0.94}$ and $ZrC_{0.89}$. This observation is puzzling and perhaps misleading. We have already noted the qualitative, if not quantitative, correspondence of Debye temperature obtained from elastic constant measurements with those obtained from low-temperature heat capacity measurements (Table II of Chapter 4). The Debye temperatures from heat capacity measurements show a strong dependence on the carbon-to-metal ratio in the monocarbide (see Fig. 2 of Chapter 4). No low-temperatures C_p measurements have been performed on ZrC_{1-x} as a function of carbon content, but the carbides NbC_{1-x} and MoC_{1-x} do show a strong dependence of Debye temperature on carbon content and a similar behavior is expected for ZrC_{1-x}. We would expect Debye temperatures of carbides derived from elastic constant data to show this same type of dependence, and therefore, we would expect the elastic constants to depend on the carbon-to-metal ratio.

There is another reason for expecting a dependence of elastic constants on the carbon-to-metal ratio. The elastic constants are an indirect measure of the interatomic bonding. The presence of carbon in the lattice promotes strong metal-to-nonmetal bonds as well as possibly enhancing the strength of the metal-to-metal bonds. Removing some of the carbon from the lattice should weaken the average bond strength and this reduction should be reflected in the elastic constants. While it is possible that the elastic constants of ZrC_{1-x} are relatively independent of the carbon-to-zirconium ratio, this single observation should not be interpreted as implying that the elastic constants of other carbide systems are not strongly dependent upon carbon concentration. Perhaps the discrepancy in values of E for TaC_{1-x} already noted in Table II is due to the different carbon contents. Until these dependencies are studied in several carbide systems, experimenters should be careful in characterizing their crystals as to carbon content, free carbon content, impurity content, and porosity. In this regard, it is difficult to compare some of the results in Table II because for several samples the carbon-to-metal ratio was unspecified.

In Table III we compare the elastic constants E and G of the carbides with those of several common elements and alloys. The most obvious feature of Table III is that the elastic constants of the carbides are about double those of the parent transition-metal element. In fact, the carbides have some of the highest values of elastic constants for all materials. Of the transition-metal elements only the elastic constants of W, Re, Os, and Ir are as high as those of the carbides.

It is interesting to speculate on the reasons for enhanced values of E

TABLE II

ELASTIC PROPERTIES OF CARBIDES AT ROOM TEMPERATURE

Carbide	Young's modulus 10^6 psi	Shear modulus 10^6 psi	Bulk modulus 10^6 psi	Poisson's ratio	Reference	Comments
$TiC_{0.91}$	65^a	27^a	35^a	0.191^a	(a)	Single crystal
TiC	39–67	16–28	—	—	(b)	Self-bonded sintered compact, quoted by Lynch et al. (b)
$ZrC_{0.94-0.89}$	59^a	25^a	32^a	0.187^a	(a)	Single crystal
$ZrC_{0.964}$	56^a	23.5^a	30^a	0.191^a	(c)	Hot-pressed 3% porosity, values adjusted to zero porosity
$HfC_{0.967}$	67^a	28^a	35^a	0.18^a	(d)	Hot-pressed, 3% porosity, values adjusted to zero porosity
HfC	46–61	26	—	—	(b)	Self-bonded sintered compact, quoted by Lynch et al. (b)
VC	63	—	—	—	(b)	Self-bonded sintered compact, quoted by Lynch et al. (b)
NbC	49	—	—	—	(b)	Self-bonded sintered compact, quoted by Lynch et al. (b)
$NbC_{0.964}$	71^a	29^a	44^a	0.23^a	(d)	Hot-pressed, 8% porosity, values adjusted to zero porosity
$NbC_{0.97}$	74^a	31^a	43^a	0.21^a	(e)	Hot-pressed, values extrapolated to zero porosity, see Fig. 2
$TaC_{0.90}$	44	17	32	—	(f)	Single crystal
$TaC_{0.994}$	78^a	31^a	50^a	0.24^a	(d)	Hot-pressed, 15% porosity, values adjusted to zero porosity

TaC	53	—	—	—	(b)	Self-bonded sintered compact quoted by Lynch et al. (b)
Cr_3C_2	56	—	—	—	(b)	Self-bonded sintered compact, quoted by Lynch et al. (b)
Mo_2C	33	—	—	—	(b)	Self-bonded sintered compact, quoted by Lynch et al. (b)
WC	97	—	—	—	(b)	Self-bonded sintered compact, quoted by Lynch et al. (b)
$WC_{1.007}$	90^a	38^a	48^a	0.185^a	(d)	Hot-pressed, 15% porosity, values adjusted to zero porosity

^{a} Preferred value.

References for Table II

a. R. Chang and L. J. Graham, *J. Appl. Phys.* **37**, 3778 (1966).
b. J. F. Lynch, C. G. Ruderer, and W. H. Duckworth, eds., "Engineering Properties of Selected Ceramic Materials". Am. Ceram. Soc., Columbus, Ohio, 1966.
c. H. L. Brown and C. P. Kempter, *Phys. Status Solidi* **18**, K21 (1966).
d. H. L. Brown, P. E. Armstrong, and C. P. Kempter, *J. Chem. Phys.* **45**, 547 (1966).
e. D. A. Speck, *Am. Ceram. Soc., 1970* (to be published); private communication (1969).
f. R. W. Bartlett and C. W. Smith, *J. Appl. Phys.* **38**, 5428 (1967).

TABLE III

COMPARISON OF THE ELASTIC MODULI OF VARIOUS COMMON MATERIALS AND THE CARBIDES

Material	Young's modulus (10^6 psi)	Shear modulus (10^6 psi)	Comments
Aluminum alloys	10	4	
Niobium	15	5	
Titanium	17	5	
Copper	18	6	
Vanadium	18	7	
Silicon	24	10	
Tantalum	27	10	
Stainless steel (18–8)	28	9	
Steel (plain carbon)	29	11	
Boron	50	—	
Tungsten	57	24	
ZrC	59	25	Single crystal
VC	63	—	Polycrystalline
B$_4$C	65	—	
TiC	65	27	Single crystal
HfC	67	28	Polycrystalline
NbC	74	31	Polycrystalline
TaC	78	31	Polycrystalline
WC	90	38	Polycrystalline
Diamond	115	—	
Graphite	144	—	

and G for the carbides over the parent transition-metal elements. We have already stated that the elastic constants are an indirect measure of the interatomic bonding and that the presence of carbon in the lattice promotes strong metal-to-nonmetal and metal-to-metal bonds. Therefore, the relatively high values of the elastic constants imply that the interatomic bonding is stronger in the carbides than in most transition elements. Not only is the bonding in carbides strong, but we can also tell something about the type of bonding involved, e.g., ionic, convalent or metallic.[1] First we can show that the bonding is *not* primarily ionic as in Me$^+$C$^-$ or Me$^-$C$^+$. For purely ionic bonding the bulk modulus K varies inversely with the fourth power of the interionic spacing R_0 by the relationship (15)

$$K = \frac{(n-1)e^2Z^2\alpha}{18R_0^4} \tag{5}$$

[1] A similar discussion of this point has been given by Y. A. Chang, L. E. Toth, and Y. S. Tyan, *Trans. AIME* (to be published).

where n is a constant about 9, e is the electron charge, Z is the number of charges transferred per ion, and α is the Madelung constant. If the values of K for a series of ionic compounds are plotted against the interionic distance R_0 on a log–log, scale, the slope should be -4. Typical plots of K versus R_0 for transition elements have higher slopes, about -6.8 (*16*).

In Fig. 3, values of K and R_0 are plotted for several carbides and transition metals for which data are available.[2] The slope of the carbide curve is similar to that for the transition elements and does not equal -4. Because the value of Z may vary from compound to compound, it is difficult to completely exclude ionic bonding from consideration. It is reasonable,

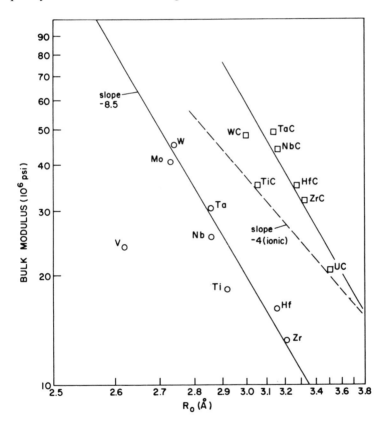

Fig. 3. When the bulk modulus K is plotted against the interionic distance R_0, the variation of K with R_0 is similar for the carbides and the transition elements. The line for the carbides is displaced toward higher K values. The variation for the carbides is different from that expected for ionic compounds in which case the K versus R_0 line should have a slope of -4.

[2] Data for UC is from C. P. Kempter, *J. Less-Common Metals* **10**, 294 (1966).

however, to assert that the bonding in carbides is similar to that in the transition elements because of the similar slopes of the lines.

The point for TiC deviates seriously from the line drawn through the other $B1$ structured carbides; the values for Ti and V, however, also deviate from the line through the transition-element points. It is not unusual to find that the transition elements and compounds for the first row have different mechanical and electrical properties from the elements of the second and third rows. The data point for WC also deviates from the line for the carbides; WC, however, has a different crystal structure from the other carbides.

Bearing in mind the limited nature of the present data, we can state that the atomic bonding is similar to that found in the transition elements. The presence of carbon in the lattice has, however, greatly stiffened the lattice. Comparison of K and R_0 values for HfC and ZrC illustrates this point. The formation of the monocarbides HfC and ZrC occur with only a slight dilation of the lattice, yet the values of K for the carbides are more than twice those for the elements. Current theories of atomic bonding in carbides, as discussed in detail in Chapter 8, indicate that carbon in the lattice creates strong metal-to-carbon bonds and also increases the strength of the metal-to-metal bonds. If this bonding hypothesis is correct, then it is not surprising to find that the slope of the K versus R_0 curve of the carbides is similar to that for the transition elements but displaced toward higher K values.

III. Plastic Deformation

At higher stresses, materials will continue to deform in either a plastic or brittle fashion depending upon the temperature. At room temperature, the carbides undergo little, if any, plastic deformation and fracture in a brittle fashion with increasing stress levels. At high temperatures the carbides become quite ductile and deform plastically. In the absence of imperfections the transition between elastic and plastic deformation should theoretically occur at the approximate stress level of either $G/30$ or $E/20$. The first limit corresponds to gross slipping of one atomic plane over the other, and the second limit corresponds to pulling apart of two adjacent atomic planes. Much lower strengths are observed, however, because of the presence of imperfections. Either dislocations move under an applied shear stress or else the material fractures in a brittle fashion because of the presence of preexisting internal or surface cracks or because of insufficient dislocation mobility to stop crack propagation. In discussing the plastic deformation of carbides, we shall first consider the deformation due to dislocation movement, second the deformation due to fracture or cleavage, and third the effect of temperature and other variables on the transition between brittle and ductile deformation.

IIIA. Slip Systems

Williams and Schaal (*17*) first investigated the slip systems for dislocation glide in TiC. Above 800°C TiC plastically deforms on {111} planes as determined by etch pit observations on single crystals. They suggested that the most likely slip direction was ⟨1T0⟩. This particular slip plane was later confirmed by Brookes (*18*) using etch pit techniques; he also found evidence for extensive plastic deformation due to dislocation movement at 1100°C. Hollox and Smallman (*19*), using contrast electron microscopy, definitely established the ⟨1T0⟩ slip direction and confirmed the {111} slip planes in TiC. This slip system is maintained at compositions deviating from the

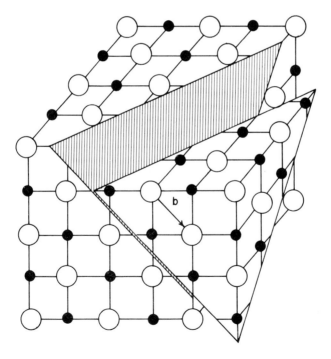

Fig. 4. The {111} ⟨1T0⟩ slip system, illustrated above, is the primary one for carbides with the *B*1 crystal structure. This slip system is also the primary one in fcc metals.

stoichiometric. Lye *et al.* (*20*) investigated the slip system in $VC_{0.84}$ and while they could not conclusively define a slip system, their results are consistent with slip on {111} planes. (Here we consider VC_{1-x} to have the $B1$ crystal structure even though the carbon atoms order. See Chapter 2.) They further pointed out that the dislocation structure and slip mechanisms of VC appear to be similar to those observed in TiC. For ZrC_{1-x} slip has been induced on the {111} $\langle 1\bar{1}0 \rangle$, {110} $\langle 1\bar{1}0 \rangle$ and {001} $\langle 1\bar{1}0 \rangle$ by controlling the orientation of a single crystal relative to the applied stress (*21*). The stress necessary to induce slip on {111} $\langle 1\bar{1}0 \rangle$ is about the same as to induce slip on {110} $\langle 1\bar{1}0 \rangle$ while slip on {001} $\langle 1\bar{1}0 \rangle$ is not favored. Slip systems in other carbides with the $B1$ structure have not been investigated. For all the investigated $B1$ carbides, however, slip probably occurs on the {111} $\langle 1\bar{1}0 \rangle$ slip system. There is also no indication that deviations from stoichiometry or impurities change the principal slip system.

The {111} $\langle 1\bar{1}0 \rangle$ slip system for carbides is illustrated in Fig. 4. The other slip systems {110} $\langle 1\bar{1}0 \rangle$ and {001} $\langle 1\bar{1}0 \rangle$, upon which deformation can occur in carbides, are illustrated in Figs. 5 and 6. Materials generally slip on close-packed planes and in close-packed directions. For the carbides the {111} planes are the closest-packed and displacements in the $\langle 1\bar{1}0 \rangle$

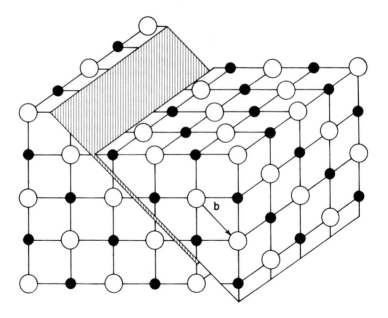

Fig. 5. Slip can also be induced on the {110} $\langle 1\bar{1}0 \rangle$ slip system in carbides with the $B1$ crystal structure. For ZrC_{1-x} this system may be the primary one. Ionic crystals with the $B1$ crystal structure prefer to slip on this system.

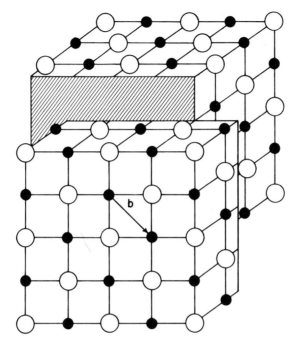

Fig. 6. Slip can be induced on the {001} ⟨1$\bar{1}$0⟩ slip system in carbides with the *B*1 crystal structure although at higher stress levels than on other systems.

directions lead to the shortest Burger's vectors. The reason for selecting the close-packed direction (shortest Burger's vector) is two-fold: the energy of a dislocation line goes as the square of the Burger's vector and the Peierls' stress to move a dislocation goes as an exponential of the Burger's vector. The {111} ⟨1$\bar{1}$0⟩ slip system is the same as commonly found for fcc metals. The probable reason for this slip system being the primary one in carbides is that since the carbon atom is small it occupies an octahedral interstitial site with only a small expansion of the metal sublattice; deformation of the monocarbides with the *B*1 structure therefore occurs on the same slip systems as in fcc.

If the carbides were ionic as in Me$^+$C$^-$ or Me$^-$C$^+$ they would not slip on {111} ⟨1$\bar{1}$0⟩. Typically ionic materials such as LiF, NaF, MgO, and NaCl which have the *B*1 crystal structure have {110} as primary glide planes (*22, 23*) and {100} as secondary glide planes (*24*). Slip on {111} does not occur normally. The reason for preferring {110} ⟨1$\bar{1}$0⟩ in ionic compounds is illustrated in Figs. 7–9. For an ionic substance, slip on {111} ⟨1$\bar{1}$0⟩ produces strong repulsive coulombic interactions across the shear plane in the

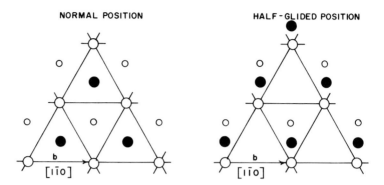

Fig. 7. In the carbides slip occurs on the {111} ⟨1̄10⟩ slip system. The unslipped or normal arrangement of {111} planes is illustrated on the left and the partially slipped arrangement on the right. If the carbides were ionic their slip on this system would produce strong Coulombic repulsive forces in the half-slipped position. Because slip does occur on this system, the carbides are probably not strongly ionic. ○ : Me ions in the A (111) plane; ○ : C ions in the X (111) plane; ● : Me ions in the B (111) plane. Sequential ordering of (111) planes in the *B*1 structure is AXBX′CX″

half-glided position (displacement = $b/2$). This repulsion is due to the close proximity of cations (anions) in the half-glided position in addition to the normal repulsive contributions in ionic bonding (Fig. 7). The same process occurs at the core of a {111} ⟨1̄10⟩ dislocation. On the {110} ⟨1̄10⟩ system the Coulombic attractive forces are maintained in the half-glided position and therefore this system is the primary glide system in ionic compounds

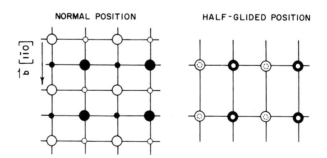

Fig. 8. Ionic compounds with the *B*1 crystal structure typically deform on {110} ⟨110⟩ slip systems. The unslipped or normal arrangement of {110} planes is illustrated on the left and the half-glided position on the right. During shear on {110} planes the attractive Coulombic forces are maintained and therefore most ionic crystals prefer to deform on these planes. ○, ○ : ions above (110) shear plane; ●, ● : ions below (110) shear plane.

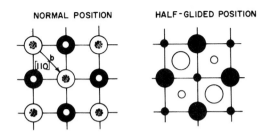

Fig. 9. Shear can be induced on {001} ⟨110⟩ slip systems in ionic materials. The critical resolved shear stress for this system is higher than for that on {110} ⟨110⟩ because in the half-glided position there is no net Coulombic attraction between adjacent planes on either side of the slip plane. ○, ○ : ions above (001) shear plane ; ●, ● : ions below (001) shear plane.

with the $B1$ crystal structure (Fig. 8) (*22*). In the half-glided position on a {100} plane the attractive Coulombic forces are destroyed because of a symmetric distribution of plus and minus charges across the shear plane (Fig. 9) (*22*). Only the normal repulsive forces in ionic bonding remain. This system, however, is still more favorable than {111} ⟨1Ī0⟩ for ionic compounds because slip on {111} ⟨1Ī0⟩ creates an additional repulsive force in the half-glided position due to the close proximity of cations (anions).

Because the carbides do slip on {111} ⟨1Ī0⟩, a strong case can be made on the basis of slip planes that the ionic contributions to the bonding in carbides are weak. This conclusion is further substantiated by theoretical band structure calculations (see Chapter 8).

Hollox and Smallman (*19*) and Williams (*25*) pointed out the importance of the {111} ⟨1Ī0⟩ slip system in the search for ductile ceramic materials. For a polycrystalline aggregate to deform in such a manner that the deformation in one grain is compatible with neighboring grains, it is theoretically necessary to have at least five independent slip systems operative. In practice, more than five systems are required for ductility. For ionic ceramic materials with the $B1$ crystal structure which slip on {110} ⟨1Ī0⟩, there are only two independent systems. For the carbides which slip on {111} ⟨1Ī0⟩ the five independent slip systems are available. Thus in commercial applications it should not be necessary to restrict certain uses to single crystals because polycrystalline carbides can be made ductile. Optimizing the ductility, however, depends upon eliminating the porosity. The pores act as stress concentrators (notch effect) and initiate cracks which cause brittleness. When the carbide is fully dense there are no pores to act as stress concentrators and the grain boundaries can inhibit the flow of dislocations and increase the yield strength. Williams (25) observed this enhancement for polycrystalline TiC. Hollox (3) reported that fully dense coarse grained TiC (2 mm grain size) has a ductility of about 30% at 1500°C and a yield strength about ten times that of single crystals. Furthermore, he proposed that decreasing the grain size without introducing porosity should further increase both the yield and fracture strength.

Slip systems in Mo_2C and WC have been investigated using visual observations of slip around Knoop and Vickers indentations at room temperature (26–31). In Mo_2C the primary slip system is (0001) $\langle 2\bar{1}\bar{1}0 \rangle$ or slip along the basal plane. Secondary slip occurs on $\{10\bar{1}0\}$ $\langle 2\bar{1}\bar{1}0 \rangle$. The twinning system is $\{10\bar{1}2\}$ [0001] (see Fig. 10) (26). In WC the slip planes are $\{1\bar{1}00\}$, and the probable directions are $\langle 0001 \rangle$ and $\langle 11\bar{2}0 \rangle$ (see Fig. 11) (27, 28). Since the c/a ratio for WC is approximately 0.976, the planes for closest packing of the tungsten atoms are $\{1\bar{1}00\}$.

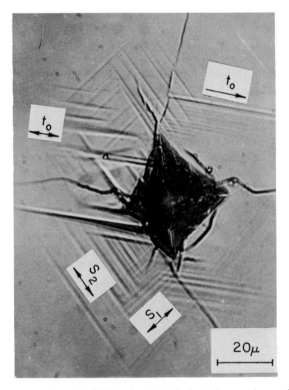

Fig. 10. A room temperature Vickers hardness indentation produces slip in a single crystal of Mo_2C. The primary slip system S_1 is (0001) $\langle 2\bar{1}\bar{1}0 \rangle$, the secondary slip system S_2 is $\{1010\}$ $\langle 2\bar{1}\bar{1}0 \rangle$, and the twinning system is $\{1012\}$ [0001]. [After Vahldiek and Mersol (26).]

Fig. 11. A microhardness indentation on a (0001) surface of WC showing slip traces produced at room temperature at approximately 1500 magnification. [After Takahashi and Freise (28).]

It is interesting to observe in Figs. 10 and 11 that slip occurs around microhardness indentations at room temperature (26–31). An interesting study of slip at room temperature around microhardness indentations in

TiC was performed by Williams (*25*). He observed that since dislocation glide is an anisotropic process the size of a microhardness indentation produced partly by slip line formation should show some orientation dependence. Using cleaved {100} surfaces and a Tukon indenter (wedge shaped) he observed that the microhardness varied nearly sinusoidally with the angle between the indenter and the [100] direction in the crystal. The amplitude of the variation was about 8% and was a minimum when the shear stress on {111} was also a minimum. Thus, even at room temperature there is evidence for some dislocation motion.

IIIB. Critical Resolved Shear Stress

The shear stress resolved along the slip plane and in the slip direction necessary to cause macroscopic slip at a given temperature in a single crystal is called the critical resolved shear stress (CRSS). The CRSS has been studied as a function of temperature and composition for TiC_{1-x} (*17, 19, 32*) and as a function of temperature for NbC (*32*), ZrC (*32*), $VC_{0.84}$ (*20*). For ZrC, the CRSS for several different slip systems have been studied as a function of temperature (*21*). These measurements have been confined to temperatures greater than 800°C, or above the brittle-to-ductile temperature. The CRSS was usually determined by using a compressive force along $\langle 100 \rangle$ directions. In some cases the CRSS were determined in three point bending tests. This technique results in CRSS values that are high by about 20% (*33*). Most recent measurements are being confined to compression tests.

Figure 12 shows the variation of the CRSS with temperature for several carbides (*20, 21, 32*). The CRSS decreases rapidly with temperature, following the equation (*32*):

$$\tau = \tau_0 \exp(-KT). \tag{6}$$

At 1800°C, NbC has the highest CRSS, 7 kg/mm² (10,000 psi), but at 1250°C $VC_{0.84}$ has the greatest value for CRSS. As Lye *et al.* (*20*) point out, the ratio of compressive yield strength to density for $VC_{0.84}$ (~400,000 in.) at 1250°C is considerably greater than that of TiC (~125,000 in.). This fact makes VC a very interesting high-strength and low-weight material up to 1200–1300°C.

In TiC_{1-x} the CRSS decreases with decreasing C/Ti ratio (Fig. 13) (*32*). This decrease in CRSS is exactly what one would expect from considerations of the number of C—Ti bonds that must be broken when a dislocation moves. Introducing vacancies on the carbon sublattice decreases the number of C—Ti bonds and hence makes dislocation motion easier. The strength of

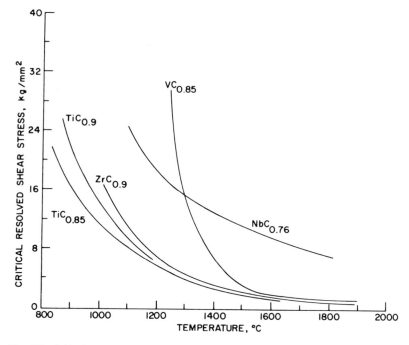

Fig. 12. Critical resolved shear stress versus temperature for several single crystal carbides. [After references (*20, 21, 32*).]

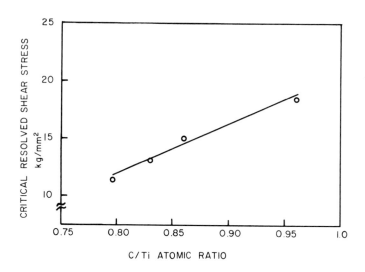

Fig. 13. At 927°C the CRSS increases with increasing carbon concentration in TiC_{1-x}. [After Williams (*32*).]

the Ti—Ti bond is also diminished in the formation of vacancies on the carbon sublattice. Both contributions, which tend to make dislocation motion easier, offset the increased resistance to dislocation motion caused by an interaction of the dislocation with the vacancies. This interaction is probably small since the vacancies are on the carbon sublattice and the net distortion of the lattice, as indicated by the variation in lattice parameter, is small.

In VC_{1-x}, there is an interesting dependence of the CRSS on the C/V ratio (3). As pointed out in Chapters 2 and 3, the carbon atoms below about 1200°C order in VC_{1-x} at the compositions V_8C_7 and V_6C_5. Cubic VC_{1-x} probably does not exist below 1200°C. The ranges of homogeneity of these ordered phases are not known. The CRSS is strongly dependent on the C/V ratio; there is a pronounced maximum in CRSS at the V_6C_5 composition (Fig. 14). The brittle-to-ductile transition temperature is also highest for this composition (3).

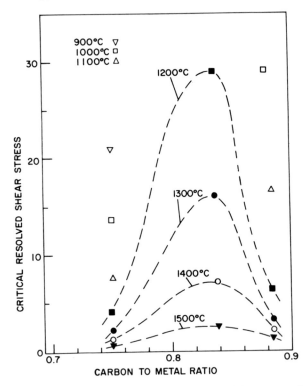

Fig. 14. There is a pronounced maximum in the CRSS of the monocarbide of vanadium. This maximum is believed to be associated with the ordering of carbon atoms. [After Hollox (3). Courtesy of the *Journal of Less-Common Metals.*]

In $ZrC_{0.90}$ the CRSS was studied as a function of temperature for several slip systems (21). Slip was induced on several systems by changing the crystal orientation such that the Schmidt factor favored slip on a particular system. Slip on $\{110\}$ planes is slightly favored over slip on $\{111\}$ planes although the difference is not exceedingly large. On the other hand, slip on $\{100\}$ planes requires a resolved shear stress nearly twice as great as the other two planes.

IIIC. Deformation Mechanisms

The carbides are brittle at room temperature for two reasons, the ease of crack nucleation and crack propagation. Surface flaws and internal pores act as crack initiation sites and the plastic flow at the tip of a crack is insufficient to absorb enough energy to stop the crack. Should either of these conditions not be present or should one factor be somehow controlled, then the low-temperature brittle behavior of the carbides could be avoided and their usefulness increased. In this section the factors governing the dislocation mobility in carbides are considered and in a later section the factors causing fracture are investigated.

The mechanisms by which dislocations move and the rate controlling factors for dislocation mobility are poorly understood in carbides. The deformation mechanism in TiC, for example, is governed by at least three activation energies depending upon the temperature range. The factors controlling dislocation mobility in carbides include a Peierls stress, the diffusion rate of carbon and the diffusion rate of the metal atoms. At low temperatures ($< 1200°C$) a Peierls stress is probably rate controlling in TiC, between about 1200 and 1500°C the diffusion rate of carbon may be rate controlling and above that temperature the diffusion rate for Ti apparently is rate controlling.

A number of investigators (17, 19, 32, 34) have suggested that the low dislocation mobility in carbides at low temperatures is due to a high Peierls stress (intrinsic lattice resistance to dislocation motion). The proportional increase in yield stress with C/Ti ratio is evidence for this interpretation although other mechanisms could produce similar results. Figure 13 shows that the CRSS increases with increasing carbon content at 927°C. Generally, deviations from stoichiometry and the generation of point defects tend to increase the yield strength, since the point defects control the dislocation mobility. Point defect–dislocation interactions are apparently not significant in titanium carbide because the vacancies exist on the carbon sublattice (19). When a dislocation moves atomic bonds must be broken (Peierls stress). In TiC_{1-x} increasing the carbon content increases the bond strength. This contribution to the resistance to dislocation motion offsets the dislocation–vacancy interaction.

Williams (32) studied the plastic deformation of TiC in terms of "dislocation dynamics" (35, 36) and found that the rate controlling process for deformation was consistent with a Peierls mechanism. In this theory, the average dislocation velocity v is related to the applied stress σ by the expression $v = (\sigma/D)^m$, where D is a constant which is a function of temperature and is characteristic of the crystal, and m is a constant, the value of which may indicate the controlling mechanism for dislocation mobility (i.e., $m \sim 1$ in covalent materials in which it is quite certain the velocity is limited by the Peierls stress). The value of m is determined by studying the strain rate ($\dot{\epsilon}$) dependence of CRSS. Williams observed a variation in CRSS of almost a factor of 10 in changing the strain rate from 4×10^{-5} to 4×10^{-2} sec^{-1} at 1400°C and from this observation he was able to deduce that m is a function of $\dot{\epsilon}$ and lies in the range $1 < m < 10$, values indicative that the Peierls stress is limiting dislocation mobility.

While there is evidence that dislocation motion at low temperatures is limited in carbides by a high Peierls stress, the actual mechanism by which a dislocation moves is poorly understood. This problem is probably one of the most important areas for future study since the inability of dislocations to move is primarily responsible for the carbides' brittle behavior at low temperatures. Understanding the mechanism for dislocation motion and the factors governing that movement may lead to control of the mechanism and ductility at lower temperatures.

There are at least three rate-controlling processes in TiC that control the deformation mechanism and the particular controlling mechanism is a function of temperature. In addition to the Peierls mechanisms, the other two involve diffusion of carbon and titanium. Rowcliffe (37) has proposed a dislocation mechanism for slip in carbides which involves the diffusion of carbon. In this mechanism the unit dislocation splits into Shockley partials and in the dislocation motion the metal atoms move through the carbon sites. For the metal atoms to move through the carbon sites it is first necessary for the carbon atoms to diffuse either to other octahedral sites or to tetrahedral sites. The mechanism is shown in Fig. 15. The Burger's vector for one unit of slip is from B_1 to B_2. This Burger's vector and slip mechanism is the same as illustrated in Fig. 4, p. 153. This slip causes a large strain normal to the slip plane. To avoid this strain the unit dislocation splits into partials B_1 to C_1 and C_1 to B_2. For the metal atom to move from B_1 to C_1 the carbon atom at C_1 must diffuse away. Thus the process involves the diffusion of carbon.

Since this mechanism involves the diffusion of carbon, the activation energy might be related to that for diffusion of carbon. Hollox (3) has

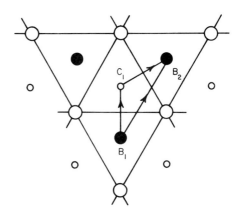

Fig. 15. Rowcliff's proposed dislocation motion in TiC. A unit of slip from B_1 to B_2 causes a high strain energy. Motion of partial dislocations from B_1 to C_1 and from C_1 to B_2 reduces this energy. The carbon atom at C must diffuse away for slip to occur. ○ : Ti atoms below shear plane; ● : Ti atoms above shear plane; ○ : C atoms below shear plane. [After Rowcliffe (*37*).]

reviewed this situation and finds that there is no simple correspondence between the activation energy for slip and the activation energy for diffusion of either C or Ti. The activation energies for slip can be determined from the variation with temperature of the CRSS and the dependence of CRSS on strain rate. Williams (*32*) who first determined the activation energy for slip in TiC_{1-x} single crystals observed that at about 1150°C the energy changed from 3.0 eV above 1150°C to 1.7–2.3 eV below 1150°C. The range of values below 1150°C corresponds to different carbon concentrations in TiC_{1-x}. The lowest value 1.7 corresponds to the composition $TiC_{0.79}$ and the highest value 2.3 corresponds to $TiC_{0.97}$. Above 1150°C, however, there appears to be no composition dependence of the activation energy. The temperature at which the activation energy changes also depends upon the nonmetal-to-metal ratio, being 1150°C for $TiC_{0.83}$ and 1305°C for $TiC_{0.95}$ (*19*, *32*).

Comparison of the activation energies for slip with those for diffusion is difficult because the activation energies for diffusion in carbides are controversial and inexactly known. From layer growth experiments, which are not accurate means of determining diffusion values, the activation energy for diffusion has been found to be for C, about 2.7 eV (*38*, *39*). The layer growth technique, however, measures an average diffusion coefficient for a range of compositions and thus, for any specific composition the above value is only approximate. Using the more accurate radio tracer technique, Sarian (*40*) found that the activation energy for carbon in TiC was about 4.0–4.5 eV. An equally high value was observed for carbon in $ZrC_{0.96}$ (*41*, *42*). The higher activation energy for carbon diffusion, 5.0 eV, does not agree with the activation energies for slip in TiC (3.0 or 2.3–1.7 eV). The lower reported activation energy for carbon diffusion, 2.7 eV, agrees with the higher activation energy for slip, 3.0 eV, but because of the inaccuracies of the layer growth technique, the radio-tracer value is preferred. Thus, there appears to be no direct correlation between the activation energy for slip and that for carbon diffusion. Exact agreement, however, is not necessary since the Rowcliffe mechanism for slip might involve diffusion of carbon along the dislocation core and the activation energy for this type of diffusion would be considerably smaller.

Above about $0.5T_m$ (1535°C) the mechanical behavior of TiC is apparently influenced by the self-diffusion of Ti. The creep rate in TiC is governed by an activation energy of between 5.0 and 7.0 eV (43), which is about the same as the estimated activation energy for self-diffusion of Ti (34).

In summary, the experimental information indicates that the deformation mechanism for TiC is governed by three activation energies: Below about 1150°C the energy varies between 1.7 and 2.3 eV and is dependent upon composition; above about 1150°C the energy is about 3.0 eV and is independent of composition, and above about $0.5T_m$ the energy for creep which probably involves dislocation climb is about 5–7 eV. Above $0.5T_m$ the energy can be correlated with the self-diffusion of Ti atoms, while below that temperature the self-diffusion of carbon is probably involved although a direct correlation cannot be made. The difference in deformation mechanism above and below about 1150°C is not clear, although the rate controlling process above 1150°C may be the diffusion of carbon and below that temperature the Peierls force may be rate controlling.

The point which should be emphasized here is that the actual mechanisms for plastic deformation in carbides are not well understood. The rate-controlling process changes with temperature three times and for each instance the exact mechanism is poorly defined. This is one area where additional research effort should be placed because understanding these mechanism may lead to control of the low-temperature brittleness of carbides.

IIID. Annealing Behavior

Hollox and Smallman (19, 44) studied the annealing behavior of TiC$_{0.97}$ by electron microscopy. The material was deformed at 1150°C and showed evidence of dislocation dipoles and elongated dislocation loops. During the annealing process, the dislocation loops pinch off into trails of smaller loops that grow and coalesce (see Figs. 16 and 17). If the specimen is annealed at 1400°C for 15 min, no dislocation loops are present; they have annealed out to form the general dislocation network of the crystal.

From the time and temperature necessary for the dislocation loops to diffuse it is possible to estimate the activation energy for diffusion of vacan-

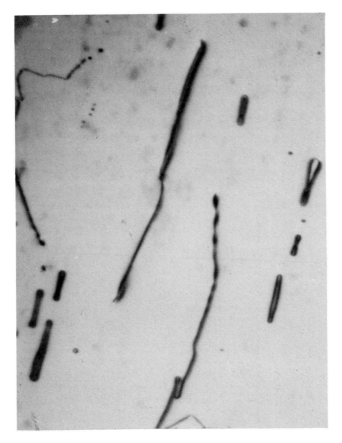

Fig. 16. Deformation of TiC produces elongated dislocations. [After Hollox and Smallman (*19, 44*).]

cies. This energy Q_D was estimated to be 5.25 eV for $TiC_{0.97}$. Subsequent studies by the same authors show that Q_D is a function of carbon composition, decreasing as carbon concentration decreases. The general mechanism for annealing thus remains the same throughout the composition range in TiC_{1-x}, but the annealing temperature becomes less as deviations from stoichiometry increase. It is interesting to note that three processes have nearly the same activation energy—diffusion of Ti, creep, and the annealing behavior (*3*). Thus it seems reasonable to assume that the creep and annealing processes are governed by the diffusion of Ti vacancies.

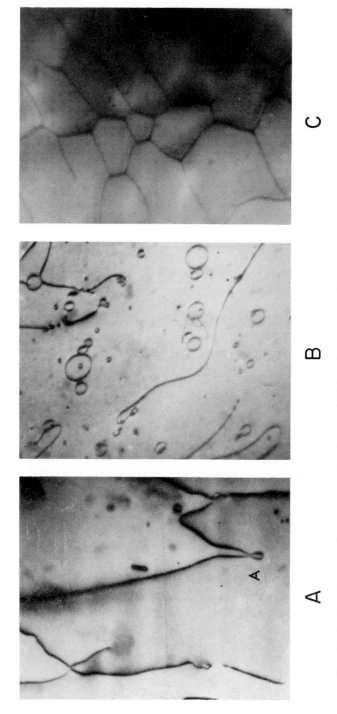

Fig. 17. During the annealing process the elongated dislocations pinch off dislocation loops as shown at "A" in (A). These loops then grow as shown in (B). During the final stages of annealing the dislocation loops anneal out and join the regular dislocation network shown in (C). [After Hollox and Smallman (*19, 44*).]

IV. Fracture and Ductile–Brittle Transition Temperature

Below about 800°C the carbides fail in a brittle fashion. This brittleness is not due to any discontinuous change in the mechanism responsible for plastic deformation but rather due to the fact that the stress required for dislocation movement is so high that the competing process of fracture takes place. The situation is probably analogous to that which causes low temperature brittleness in bcc metals. The transition from a ductile-to-brittle behavior occurs because of the strong temperature sensitivity of the CRSS and the relative temperature insensitivity of the fracture stress (45). The strong temperature sensitivity of the CRSS in carbides has been experimentally confirmed (see Section IIIB) but the temperature dependence of the fracture stress has not yet been studied. The competition between plastic deformation and fracture is illustrated in Figs. 18 and 19 where σ_f is the fracture strength or cleavage strength, σ_y is the yield strength, and T_{DB} is the ductile-to-brittle transition temperature. Below T_{DB} the carbide fractures or cleaves because the stress necessary to produce fracture σ_f is lower than the stress needed to move dislocations σ_y; above T_{DB} the opposite is true

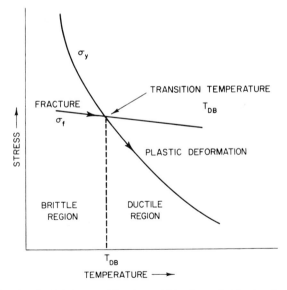

Fig. 18. The ductile–brittle transition in carbides is probably due to a competition between the stress to cause fracture σ_f and the stress to cause plastic deformation σ_y Below T_{DB}, σ_f is less than σ_y and the carbide fractures in a brittle manner; above T_{DB}. σ_y is less than σ_f and plastic deformation occurs due to dislocation motion. The transition occurs because of the temperature sensitivity of σ_y.

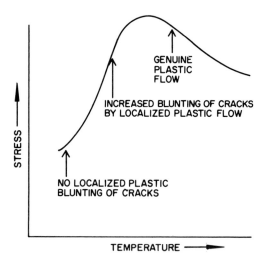

Fig. 19. In materials undergoing a brittle-to-ductile transition, the stress necessary to cause permanent deformation goes through a maximum because dislocation movements locally blunt crack propagation while at higher temperatures and stress true plastic flow occurs.

and the carbide is ductile. The yield stress has the same temperature dependence as the CRSS as the two parameters are related by the Schmidt factor.

A number of factors will influence the ductile-to-brittle transition temperature and the stress at the transition point. These include factors which affect σ_f such as surface and internal flaws and those affecting σ_y such as triaxial state of stress and the strain rate. The temperature dependence of both σ_y and σ_f play an important role in the transition. For nearly all metals the yield stress decreases with increasing temperatures (45). For fcc metals the temperature dependence of σ_y is relatively small with σ_y increasing from room temperature to liquid nitrogen temperatures by a factor of only 2 (45). No ductile-to-brittle transition is observed. For bcc metals the yiled stress in the same temperature interval increases by a factor of 3 to 8 (45) and a transition is observed. For the carbides the temperature dependence of σ_y is comparable to the bcc metals and a low temperature brittle behavior is observed. Thus it is interesting to observe that while the carbides have the same slip systems as fcc metals, their temperature dependence of yield stress is more comparable to bcc metals or Ge.

Single crystals of carbides (TiC) fail by cleavage on {100} planes (8), but little is known about the detailed mechanism of fracture or the energy required for cleavage. The fact that carbides do cleave on {100} planes is interesting because of the common planes that cleavage could occur on

($\{111\}$ $\{110\}$ and $\{100\}$), the $\{100\}$ planes are the ones with the smallest ratio of nearest-neighbor metal-to-metal bonds to nearest-neighbor metal-to-nonmetal bonds broken per unit area. Thus cleavage on $\{100\}$ planes results in a higher proportion of broken nonmetal-to-metal bonds. This behavior would indicate that metal-metal atomic bonds are stronger than metal-to-carbon bonds. This hypothesis is in agreement with several current theories of bonding, but contrary to those relying upon the strong nonmetal-to-metal concept (see Chapter 8 for a discussion of bonding in carbides). Hollox, in a private communication, states that he has also observed cleavage on $\{110\}$ in $TiC_{0.97}$.

The fracture or cleavage stress σ_f can be significantly increased by removing surface flaws. Williams (*17, 25*) determined transverse rupture strengths of single crystal and polycrystalline TiC at room temperature. Some of the samples were tested in the as-cleaved condition while others were tested after electropolishing. The transverse rupture strengths of polycrystalline TiC were not noticeably increased by electropolishing and remained at a low value of 11,000–18,000 psi. The strengths of the as-cleaved single crystals were appreciably higher, 41,000–97,000 psi, and electropolishing increased these values to 100,000–800,000 psi. A factor of over 8 was achieved by electropolishing and removing the surface flaws. The value of 800,000 psi obtained for one electropolished TiC crystal is only about three times lower than the theoretical strength $E/20$, and it illustrates the great intrinsic strength of carbides.

The fracture strength must also be dependent upon the nonmetal-to-metal ratio. Deviations from stoichiometry will decrease the number of nonmetal-to-metal bonds that must be broken during fracture or cleavage. It is also possible that the preferred cleavage planes might change with the nonmetal-to-metal ratio but no experimental evidence is available on this point. We expect, however, that σ_f increases as stoichiometry increases and as the vacancy concentration decreases.

Increasing σ_f, as by removing surface flaws or by decreasing the vacancy concentration should, according to Fig. 18, decrease the ductile-to-brittle transition temperature. The situation in which the nonmetal-to-metal ratio is changed is more complex, however, because as σ_f is increased with increasing stoichiometry so is the CRSS altered (see Figs. 14 and 15) and therefore T_{DB} might not be altered significantly or it might even increase. This situation has not, however, been experimentally investigated.

The yield stress, σ_y is sensitive to the strain rate and any flaws or notches which will produce a triaxial state of stress. To see the dependence of σ_y on the strain rate $\dot{\epsilon}$ we can write

$$\dot{\epsilon} = nbv \qquad (7)$$

where n is the number of dislocations, b is their Burger's vector and v is the average velocity of the dislocation. The velocity depends empirically upon the stress level σ by the relationship

$$v = (\sigma/\sigma_0)^m, \tag{8}$$

where σ_0 is the stress necessary to produce a unit velocity and m is the exponent previously described in Section III. For the carbides m is relatively low ($1 < m < 10$) (32) and, hence, the velocity of the dislocations is relatively insensitive to the stress. A comparison of m values for several materials, as compiled by Harrod and Fleischer (34), is given in Table IV.

TABLE IV

COMPARISON OF m VALUES FOR SEVERAL
MATERIALS [a]

Material	m	Temperature range
Covalent crystals (Ge, Si)	1–2	$>0.3T_m$
TiC	1–10	$>0.3T_m$
bcc Metals (W, Mo)	5–10	$<0.25T_m$
LiF	~25	$<0.25T_m$
Fe–3% Si	~40	$<0.25T_m$

[a] After Harrod and Fleischer (34).

Combining Eqs. (1) and (2) and assuming an exponential temperature dependence for v ($v = Ce^{-E/kT}$ where C is a constant) leads to the equation

$$\dot{\epsilon} = nb(\sigma/\sigma_0)^m Ce^{-E/kT}. \tag{9}$$

Solving for the stress needed for a particular $\dot{\epsilon}$ yields

$$\sigma = [\sigma_0/(nbC)^{1/m}](\dot{\epsilon})^{1/m}e^{E/mkT}. \tag{10}$$

Considering the case when $m = 1$, the yield stress σ is directly proportional to $\dot{\epsilon}$. When m is large, however, the strain rate dependence of σ is small. For TiC and presumably the other carbides there is a significant dependence of the flow stress or CRSS on $\dot{\epsilon}$ (see Fig. 20). This dependence has the effect of increasing the brittle-to-ductile transition temperature because a large increase in flow stress is needed to change the dislocation velocity to accommodate any increase in $\dot{\epsilon}$. This dependence is illustrated in Fig. 21.

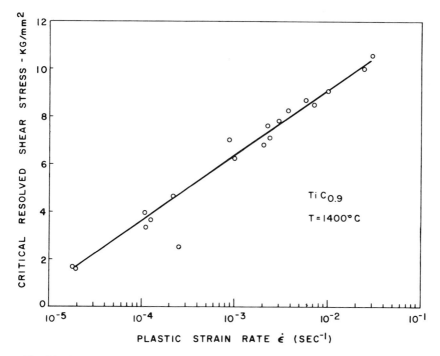

Fig. 20. In TiC$_{0.9}$ the CRSS is highly sensitive to the strain rate $\dot{\epsilon}$ indicating a small *m* value in Eq. (7). [After Williams (*32*).]

The presence of notches in the sample, as would be found in poly-crystalline sintered carbides, produces a triaxial stress state and its effect is similar to that of increasing the strain rate. The flow stress curve in Fig. 18 is shifted to the right and higher T_{DB}'s result.

For single crystals of carbides T_{DB} is about $0.3T_m$. Approximate T_{DB} values are 800°C for TiC$_{0.95}$ (*19*, *32*), 1200°C for VC$_{0.84}$ (*20*), 900°C for ZrC$_{0.9}$ (*32*), and 1000°C for NbC$_{0.76}$ (*32*). For hot-pressed polycrystalline carbides T_{DB} is several hundred degrees higher: for TiC$_{0.745}$ (1530°C), VC$_{0.61}$ (1230°C), NbC$_{0.95}$ (1480°C), TaC (1730°C), and WC (1320°C) (*46*).

The temperature T_{DB} may be lowered by increasing deviations from stoichiometry in TiC$_{1-x}$. Hollox and Smallman (*19*) reported that the transition in TiC$_{0.97}$ occurred at about 900°C, while Williams (*32*) reported a transition temperature of 800°C for TiC$_{0.95}$. The effect of impurities on T_{DB} in carbides is unknown. Impurities often tend to increase this transition temperature in metals especially bcc transition elements. Since substoichio-metric TiC$_{1-x}$ is often contaminated by oxygen during preparation, it may prove difficult to separate the effects on T_{DB} of stoichiometry deviations

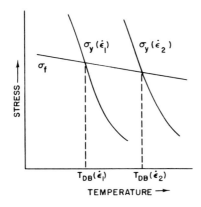

Fig. 21. The ductile–brittle transition temperature is sensitive to the strain rate $\dot{\epsilon}$. Increasing the strain rate from $\dot{\epsilon}_1$ to $\dot{\epsilon}_2$ increases T_{DB}.

from those of oxygen. Nevertheless, preparation of oxygen-free sub-stoichiometric TiC_{1-x} crystals may provide materials that possess low T_{DB} values and high strengths. In polycrystalline samples, the impurities may segregate at grain boundaries and cause brittleness.

V. Strengthening Mechanisms

Because of their strength and deformation characteristics analogous to fcc metals, carbides are potentially important materials for high-temperature structural applications. The CRSS of unalloyed carbides, however, rapidly decreases with increasing temperature. The strength of single-crystal TiC at 1600°C is only about 1/400 of its value at room temperature. A number of mechanisms, however, can be used to increase this high-temperature strength.

Williams (47) found that the CRSS of TiC crytals is increased by a factor of 5 at 1600°C by addition of a few tenths of 1% boron (see Fig. 22). He observed boride precipitates and suggested that precipitation was the strengthening mechanism. Venables (48) studied the nucleation of these boride precipitates. The precipitates of TiB_2 are approximately 10 Å thick and nucleate heterogeneously at extrinsic dislocation nodes. The platelets form a three-dimensional array by nucleating on different $\{111\}$ planes. A precipitate network is shown in Fig. 23.

Since the precipitates form at dislocation nodes, the high-temperature mechanical properties depend on the array of dislocation nodes prior to precipitation. By controlling the morphology of the dislocation nodes, boride precipitation and hence high-temperature mechanical properties may possibly be controlled (48).

Boride precipitates have also been observed in the V–C–B systems with a

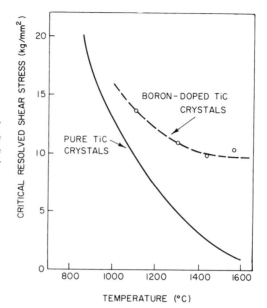

Fig. 22. Addition of a few tenths of 1% boron to TiC single crystals greatly improve the high-temperature strength. [After Williams (47).]

corresponding enhancement of the strength above that of VC (49). These precipitates have been observed on both $\{111\}$ and $\{100\}$ planes; the composition and structure of the boride have not been conclusively determined. Hollox, in a private communication, states that he has recently observed that the borides precipitate as coherent spherical particles.

Other carbides which have analogous borides should also allow precipitation hardening. An inspection of the ternary phase diagrams of a refractory transition metal with boron and carbon indicates that other likely systems are Zr–C–B, Hf–C–B, Nb–C–B, and Ta–C–B (50, 51).

Lye et al. (20) suggest another strengthening mechanism in $VC_{0.84}$ which may influence its high-temperature strength. A domain structure forms in $VC_{0.84}$ as a result of the ordering of carbon atoms in the carbon sublattice (see Fig. 24), and the domain walls may act as barriers to dislocation motion (52). The rapid decrease in the CRSS of $VC_{0.84}$ at 1250–1300°C may be due to the motion of these barriers, or due to their disappearance because of an order–disorder reaction in this temperature range, or possibly due to a change in the mechanism of carbon diffusion in the ordered compound. Such order–disorder reactions are common in the subcarbides (see Chapter 2).

Alloys formed between TiC and VC show far greater strengths at 1800°C than either parent carbide (see Fig. 25) (3). Their strengths (30,000 psi) at 1800°C and their light weight make these alloys promising high-temperature

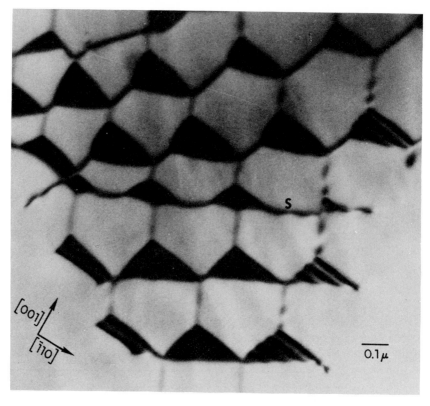

Fig. 23. Addition of boron to TiC results in the precipitation of TiB₂ platelets in {111} planes. The precipitates nucleate heterogeneously at extended dislocation nodes to form the three dimensional array shown above. [After Venables (*48*).]

structural materials. A VC–25 at. % TiC alloy also shows some indication of ductility ($\sim 1\%$) at 1000°C. The alloys appear to be two phase, one of which is ordered. A fine precipitation of the hard-ordered phase in a ductile matrix may be responsible for the improved strength characteristics.

VI. Microhardness

The carbides are used extensively in commercial applications such as cutting tools and wear-resistant surfaces because of their great hardness. In most of these applications the carbide powders are bonded with the aid of

Fig. 24. The excellent high strength properties of $VC_{0.84}$ may be due to the domain structure shown above. The domains result from an ordering of the carbon atoms. The domain structure is revealed with polarized light between crossed nicols on a $\langle 100 \rangle$ cleavage plane. [After Venables *et al.* (*52*).]

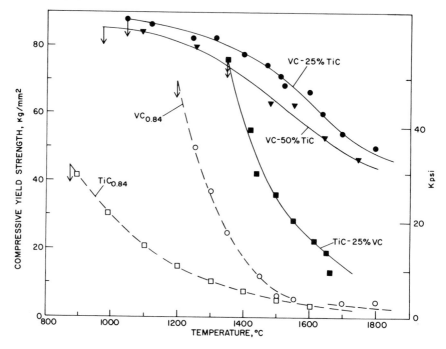

Fig. 25. Alloys of VC–TiC show improved high temperature strengths. [After Hollox (*3*).]

small amounts of cobalt. The cobalt increases the toughness and impact resistance of the carbide. The most widely used material is tungsten carbide. It has excellent room-temperature and high-temperature hardness and it is also one of the most inexpensive carbides to manufacture.

In view of the importance of carbides in commercial applications because of their hardness, it is surprising to find that not all the variables affecting hardness have been thoroughly studied. The hardness will be seriously affected by temperature, nonmetal-to-metal ratio, alloying behavior, defect structure and precipitate structure. Nearly all hardness measurements have been performed on polycrystalline samples. Because a hardness test is performed on a single grain, the hardness values reported here should reflect true bulk properties and not reflect grain boundary influences.

Westbrook (*1*) performed extensive measurements on the microhardness of carbides as a function of temperature. A summary of these data is shown in Fig. 26. At room temperature TiC is the hardest carbide ($\sim$ 3000 kg/mm²), but it rapidly loses its hardness with increasing temperature. WC is the hardest carbide at elevated temperatures: $\sim$ 1000 kg/mm² at 1000°C.

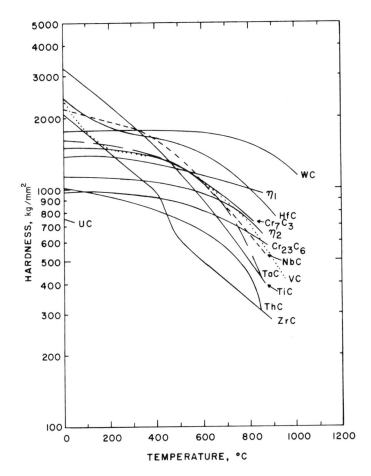

Fig. 26. The microhardness of carbides decreases with increasing temperature. At room temperature TiC is the hardest binary carbide but at 1000°C, WC is the hardest binary carbide. [After Westbrook and Stover (1).]

The hardness is a sensitive function of the carbon-to-metal ratio in the monocarbides. In TiC_{1-x} the hardness increases sharply as stoichiometry is approached (see Fig. 27) (53, 54). This behavior is similar to increasing resistance to plastic deformation (CRSS) with increasing carbon content in TiC_{1-x} as previously discussed. In TaC_{1-x}, however, the opposite trend is observed: The hardness increases with increasing deviations from stoichiometry in the composition interval $TaC–TaC_{0.85}$ (see Fig. 28) (55, 56). At lower carbon concentrations, the hardness decreases (55). The values for the transverse rupture strengths as a function of carbon content in TaC_{1-x}

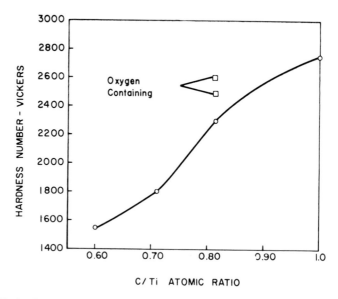

Fig. 27. In TiC_{1-x} the microhardness increases with increasing carbon content. Samples deliberately doped with oxygen have a greater hardness than oxygen-free samples. [After Cadoff *et al.* (*54*).]

do not increase as the hardness increases but rather decreases to a minimum at a composition of $TaC_{0.9}$ (*57*). It is unusual to find that the strength and hardness vary in opposite directions with composition and this situation casts doubt on some of the experimental results. The strength and hardness values for the TaC_{1-x} samples were determined either on cobalt bonded sintered compacts (*55, 56*) or carburized wires (*57*) and therefore the strengths,

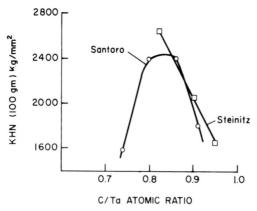

Fig. 28. In TaC_{1-x} the hardness first increases with increasing deviations from stoichiometry and then decreases (*55, 56*). The behavior may be associated with the electronic structure of TaC_{1-x}.

in particular, may not reflect the actual behavior in single crystals or fully dense carbides. The hardness values, however, probably do reflect the behavior in single crystals or fully dense materials.

Santoro (55) has suggested that the maximum in the hardness values may somehow be associated with the electronic structure of TaC_{1-x}. He found, for example, a color change and a minimum in the magnetic susceptibility at the composition $TaC_{\sim 0.83}$ where the maximum in the hardness occurs. He was not able to develop his hypothesis further because of insufficient data on the electronic structure of the carbides.

As we shall discuss further in Chapters 6–8, there is now sufficient evidence to suggest that carbide formation results in a pronounced separation of the bonding and antibonding parts of the d band. The bonding portion of the resultant band structure is filled with about 8–8.5 valence electrons; that is the bonding part is nearly filled at the composition TiC (8 electrons) while there would be about one electron in the antibonding band in stoichiometric TaC (9 electrons). Deviations from stoichiometry in TiC_{1-x} results in a depletion of electrons in the bonding portion of the band and hence lower bond strength. Thus, the hardness decreases with increasing deviations from stoichiometry (Fig. 27). In TaC_{1-x}, deviations from stoichiometry deplete electrons from the antibonding portion of the band and the average bond strength increases. Thus the hardness increases with deviations from stoichiometry in TaC_{1-x} (Fig. 28). Electrical properties such as coefficients of the electronic specific heat and magnetic susceptibilities confirm the splitting of the bands. Of course, the bonding model presented here is an oversimplification because significant deviations from stoichiometry will not only alter the position of the Fermi level in the band but also change the shape of the band.

Takahashi and Freise (28) found considerable anisotropy in the hardness values for different crystallographic planes in single crystals of hexagonal WC. On the basal plane (0001), the Vicker's microhardness was $2100 \pm 40 \text{ kg/mm}^2$, while it was only $1080 \pm 50 \text{ kg/mm}^2$ on the $(1\bar{1}00)$ plane and $1060 \pm 20 \text{ kg/mm}^2$ on the $(1\bar{1}01)$ plane.

Vahldiek et al. (58) studied the microhardness of Mo_2C and MoC as a function of crystal orientation, annealing temperatures, and load. The representative Knoop hardness for the basal plane of Mo_2C is $1360 \pm 50 \text{ kg/mm}^2$ for loads between 400 and 1000 gm. These investigators also observed an interesting variation in hardness with annealing temperatures. Hardness decreased with increasing annealing temperatures up to 2000°C, but annealing at higher temperatures produced an increase in hardness. This increase results from the precipitation of Mo and γ'-MoC from the Mo_2C matrix and the minute dispersion of these second phases in a veining substructure, in some cases Widmanstätten-type precipitates.

VII. Summary

During the past five years, considerable progress has been made in understanding the mechanical behavior of carbides. These very strong materials deform on slip systems analogous to fcc metals, but they have a temperature dependence of the flow stress analogous to bcc metals. At low temperatures their usefulness is limited by their brittleness, but at high temperatures they become quite ductile. Significant improvements in their high-temperature strengths have been achieved by controlled precipitation of borides and by alloying. The alloy VC–25 at. % TiC is particularly promising for technical applications because it retains its strength at high temperatures and has a low density.

The fact that these materials deform as fcc metals has important technological implications. A minimum of five independent slip systems is necessary for ductility in polycrystalline aggregates. Since the cubic carbides slip on $\{111\} \langle 1\bar{1}0 \rangle$ there are five independent systems available; therefore, polycrystalline aggregates can be ductile at temperatures not appreciably greater than the brittle-to-ductile transition temperature of the single crystals. Most ceramic materials do not meet this requirement until slip systems other than the primary system are activated.

It is likely that fully dense single crystals or polycrystalline carbides will be necessary for technological applications since the yield strength of ceramics decreases exponentially with porosity. The pores act as crack-propagating sites at low temperatures and prevent grain-boundary rotation at high temperatures. With fully dense coarse-grained TiC, Hollox (3) finds a 30% ductility at 1500°C and a yield strength ten times greater than that of a single crystal of the same composition. Reducing the grain size increases the yield strength. Because of this ductility and strength, fine-grained fully dense carbides will be important high-temperature structural materials in the future (3).

While significant progress has been made in improving the high-temperature strength of carbides by alloying, little has been done to reduce the brittle-to-ductile transition temperature. Their great room-temperature strengths cannot be fully utilized because of brittleness. The factors affecting this brittleness, such as stoichiometry and impurity levels, need to be more closely investigated.

References

1. J. H. Westbrook and E. R. Stover, *in* "High-Temperature Materials and Technology" (I. E. Campbell and E. M. Sherwood, eds.), p. 312. Wiley, New York, 1967.

2. J. F. Lynch, C. G. Ruderer, and W. H. Duckworth, eds., "Engineering Properties of Selected Ceramic Materials." Am. Ceram. Soc., Columbus, Ohio, 1966.
3. G. E. Hollox, *Mater. Sci. Eng.* **3**, 121 (1968/1969); see also *Nat. Bur. Stand. (U.S.)*, *Spec. Publ.* **303**, 201 (1969).
4. D. A. Speck and B. R. Miccioli, "Advanced Ceramic Systems for Rocket Nozzle Applications," Carborundum Company Report, October, 1968.
5. O. L. Anderson, *Phys. Rev.* **144**, 1553 (1966).
6. R. F. Brenton, C. R. Saunders, and C. P. Kempter, *J. Less-Common Metals* **19**, 273 (1969).
7. G. V. Samsonov, "High Temperature Materials, Properties Index." Plenum Press, New York, 1964.
8. J. J. Gilman and B. W. Roberts, *J. Appl. Phys.* **32**, 1405 (1961).
9. R. Chang and L. J. Graham, *J. Appl. Phys.* **37**, 3778 (1966).
10. B. T. Bernstein, unpublished data, reported by Chang and Graham (9).
11. J. deKlerk, *Rev. Sci. Instr.* **36**, 1540 (1965).
12. R. W. Bartlett and C. W. Smith, *J. Appl. Phys.* **38**, 5428 (1967).
13. H. L. Brown and C. P. Kempter, *Phys. Status Solidi* **18**, K21 (1966).
14. H. L. Brown, P. E. Armstrong, and C. P. Kempter, *J. Chem. Phys.* **45**, 547 (1966).
15. C. Kittel, "Introduction to Solid State Physics," 2nd ed., p. 79. Wiley, New York, 1956.
16. J. J. Gilman, "High-Strength Materials of the Future," G.E. Rep. GO-RL-2579M (1960).
17. W. S. Williams and R. D. Schaal, *J. Appl. Phys.* **33**, 955 (1962).
18. C. A. Brookes, *in* "Special Ceramics" (P. Popper, ed.), p. 221. Academic Press, New York, 1963.
19. G. E. Hollox and R. E. Smallman, *J. Appl. Phys.* **37**, 818 (1966).
20. R. G. Lye, G. E. Hollox, and J. D. Venables *in* "Anisotropy in Single-Crystal Refractory Compounds" (F. W. Vahldiek and S. A. Mersol, eds.), Vol. 2, p. 445. Plenum Press, New York, 1968.
21. D. W. Lee and J. S. Haggerty, quoted by Hollox (3).
22. M. J. Buerger, *Amer. Mineral.* **15**, 174 and 226 (1930).
23. H. B. Huntington, J. E. Dickey, and R. Thomson, *Phys. Rev.* **100**, 1117 (1955).
24. J. J. Gilman, *Acta Met.* **7**, 608 (1959).
25. W. S. Williams *in* "Propriéte's Thermodynamiques Physiques et Structurales des Dérive's Semi-Métalliques," p. 181. *Colloq. Int. Cent. Nat. Rech. Sci. No. 157 Paris*, 1967.
26. F. W. Vahldiek and S. A. Mersol, *in* "Anisotropy in Single-Crystal Refractory Compounds" (F. W. Vahldiek and S. A. Mersol, eds.), Vol. I, p. 199. Plenum Press, New York, 1968.
27. N. French and A. Thomas, *Trans. AIME* **233**, 950 (1965).
28. T. Takahashi and E. J. Freise, *Phil. Mag.* [8] **12**, 1 (1965).
29. J. Corteville, J. S. Monier, and L. Pons, *C. R. Acad. Sci.* **260**, 2773 (1965).
30. J. Corteville and L. Pons, *C. R. Acad. Sci.* **257**, 1915 (1963).
31. J. Corteville and L. Pons, *C. R. Acad. Sci.* **258**, 2058 (1964).
32. W. S. Williams, *J. Appl. Phys.* **35**, 1329 (1964).
33. W. T. Read, *Acta Met.* **5**, 83 (1957).
34. D. L. Harrod and L. R. Fleischer, *in* "Anisotropy in Single-Crystal Refractory Compounds" (F. W. Vahldiek and S. A. Mersol, eds.), Vol. I, p. 341. Plenum Press, New York, 1968.

35. For example, see W. G. Johnston, *J. Appl. Phys.* **33**, 2716 (1961); and G. T. Hahn, *Acta Met.* **10**, 727 (1962).
36. P. Haasen, *in* "Dislocation Dynamics" (A. R. Rosenfield, *et al.*, eds.), p. 701. McGraw-Hill, New York, 1968.
37. D. J. Rowcliffe, Ph.D. Thesis, University of Cambridge, 1965.
38. C. A. Vansant and W. C. Phelps, Jr., *Trans. Amer. Soc. Metals* **59**, 105 (1966).
39. L. M. Adelsberg and L. H. Cadoff, *Trans. AIME* **239**, 933 (1967).
40. S. Sarian, *J. Appl. Phys.* **39**, 3305 (1968).
41. S. Sarian and J. M. Criscione, *J. Appl. Phys.* **38**, 1794 (1967).
42. R. A. Andrievskii, V. N. Zagryazkin, and G. Ya. Meshcheryakov, *Symp. Thermodynamics, with Emphasis on Nuclear Materials and Atomic Transport in Solids, Vienna,* 1965, Vol. II, p. 172. IAEA, Vienna, 1966.
43. F. Keihn and R. Kebler, *J. Less-Common Metals* **6**, 484 (1964).
44. G. E. Hollox and R. E. Smallman, *Proc. Brit. Ceram. Soc.* **1**, 211 (1964).
45. G. E. Dieter, "Mechanical Metallurgy." McGraw-Hill, New York, 1961.
46. A. Kelly and D. J. Rowcliffe, *J. Am. Ceram. Soc.* **50**, 253 (1967).
47. W. S. Williams, *Trans. AIME* **236**, 211 (1966).
48. J. D. Venables, *Phil. Mag.* [8] **16**, 873 (1967).
49. G. E. Hollox and J. D. Venables, *Trans. Inst. Metals, Jap.* Suppl. **9**, 295 (1968).
50. H. Nowotny, E. Rudy, and F. Benesovsky, *Monatsh. Chem.* **92**, 393 (1961).
51. E. Rudy, F. Benesovsky, and L. Toth, *Z. Metallk.* **54**, 345 (1963).
52. J. D. Venables, D. Kahn, and R. G. Lye, *Phil. Mag.* [8] **18**, 177 (1968).
53. A. E. Kovalskii and T. G. Makarenko, *Zh. Tekh. Fiz.* **23**, 265 (1953).
54. I. Cadoff, J. P. Neisen, and E. Miller, *Plansee Proc., Pap. Plansee Semin. "De Re Metal."* 2*nd*, 1955, p. 50 (1956).
55. G. Santoro, *Trans. AIME* **227**, 1361 (1963).
56. R. Steinitz, *in* "Nuclear Applications of Non-fissionable Ceramics" (A. Boltax and J. H. Handwerk, eds.), p. 75. Am. Nucl. Soc., Hinsdale, Illinois, 1966.
57. H. A. Johansen and J. G. Cleary, *J. Electrochem. Soc.* **113**, 378 (1966).
58. F. W. Vahldiek, S. A. Mersol, and C. T. Lynch, AFML-TR-66-268 (1966).

6

Electrical and Magnetic Properties

I. Introduction

Transition-metal carbides and nitrides have electrical and magnetic properties similar to those of the transition-metal elements. Their properties are typically metallic and the values of parameters such as electrical resistivity, Hall coefficient, and magnetic susceptibility are comparable to those of the transition-metal elements and alloys. After a review of these parameters in this chapter several simple correlations with electron concentration are discussed. These correlations suggest that nearly stoichiometric carbides and nitrides can be treated as isoelectronic phases. The variation in properties with electron concentration is approximately accounted for by adjusting the height of the Fermi level in a supposedly rigid band. We attempt, in the following pages, to evaluate critically these correlations and to place limitations on their usefulness. In Chapter 8 on bonding and band structure, implications of these correlations will be discussed further.

Our knowledge of electrical and magnetic properties of carbides and nitrides is very incomplete. Only a few properties have been carefully measured. Most experiments have been conducted on poorly characterized sintered compacts. Since the importance of impurities, stoichiometry deviations, and porosity is often unknown, it is difficult to evaluate critically much of this data.

II. Electrical Resistivities

Most carbides and nitrides are metallic conductors with room-temperature resistivities ρ in the range 7–250 $\mu\Omega$-cm. Resistivity values have been compiled by Samsonov (1), Rudy and Benesovsky (2), and Costa and Conte (3); representative values are listed in Table I, Unfortunately, these values are imperfectly known, and large differences appear in reported values. For example, values for TiC range from 35 to 250 $\mu\Omega$-cm. The older values, (4, 5) generally higher than the more recent ones, are probably not as reliable and are not listed in Table I.

The difference in the resistivity values can usually be attributed to the following factors:

(1) Different nonmetal-to-metal ratios for supposedly the same reported composition (usually stoichiometric).

(2) Differing amounts of impurities, particularly other interstitial elements such as oxygen.

(3) Differing residual porosities and differing empirical formulas for correcting this porosity.

Deviations from stoichiometry rapidly increase the resistivity in TaC_{1-x} and TiC_{1-x} (6–8) (see Fig. 1). This increase in resistivity with decreasing carbon content is greater than would be expected from electron scattering by vacant carbon sites. If the increase were due solely to scattering by vacancies, the scattering power of a vacancy would be about ten times as great as for a vacancy in copper (8). Williams (8) has suggested that the rapid increase in ρ in TiC is due to a small number of conduction electrons, a large effective charge on vacancies, and poor screening of that charge. Lye (9) has suggested that this increase in ρ is due to a rapidly increasing density of states with decreasing carbon content. Magnetic susceptibility measurements (10) do indicate that the density of states rapidly increases with decreasing carbon content in TiC. The same explanation cannot account for the resistivity increase in substoichiometric TaC_{1-x}, since the density of states apparently decreases with decreasing carbon content (10). Since the dependence of ρ on composition is similar in TiC_{1-x} and TaC_{1-x}, it seems that several complex and as yet unresolved factors are responsible for these increases.

Studies are lacking on the effect of nonmetal impurities, particularly oxygen, on electrical resistivity. Presumably oxygen would be ionically charged in the lattice and would, therefore, effectively scatter electrons. It is especially difficult to obtain fourth-group nonstoichiometric carbides and

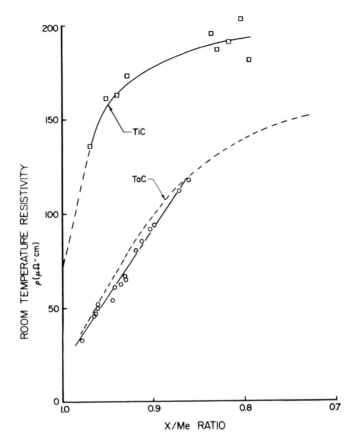

Fig. 1. The electrical resistivity rapidly increases with decreasing X/Me ratio in TiC_{1-x} and TaC_{1-x}. - - - - : TaC data from Santoro and Dolloff (7); ○ : TaC data from Cooper and Hansler (6); □ : TiC data from Williams (8).

nitrides free from oxygen contamination due to the tremendous chemical affinity of this group for oxygen. The presence of oxygen in the TiC samples, for example, could be responsible for part of the increase in ρ with stoichiometry deviations.

Most carbide and nitride samples for electrical resistivity measurements have been prepared by powder metallurgy techniques. Their high melting points and relatively slow diffusion rates make difficult the preparation of fully dense compacts. Experiments on electrical resistivities have been performed on samples containing as much as 50% residual porosity. Correcting for this porosity is a major experimental problem, since both distribution

TABLE I

ELECTRICAL RESISTIVITIES OF CARBIDES AND NITRIDES

Composition	ρ ($\mu\Omega$-cm) (room temp.)	ρ (77°K)	ρ (4.2°K)	Reference
TiC	68[a]			(a)
	51[a]			(b)
	52.5[a]			(c)
	68.2[a]			(d)
	110[a, b]			(e)
	138–188[a]			(f)
ZrC	42			(a)
	50			(c)
HfC	37			(a)
	45			(c)
VC[c]	60			(a)
	65			(c)
$VC_{0.88}$[d]	49.55	9.33	8.16	(g)
$VC_{0.83}$[d]	29.65	15.20	14.22	(g)
NbC	35			(a)
	48			(b)
	51.1			(c)
	—		35.1	(h)
$NbC_{0.975}$[d]	27.7	16.03	15.01	(g)
TaC	25[a]			(a)
	20[a]			(b)
	42.1[a]			(c)
	—		12.7	(h)
MoC	—		77	(i)
Mo_3C_2 (hex)	—		96	(i)
WC (hex)	22			(a)
TiN	25			(c)
	53.9	24.34	23.22	(g)
$TiN_{0.765}$	50.6			(c)
ZrN	21.1			(c)
	7.0	2.50	2.24	(g)
$ZrN_{0.879}$	37			(c)
HfN	33			(c)
VN	85			(c)
$NbN_{0.90}$[d]	44.35	44.95	45.10	(g)
	77 6	78.3	78.4	(g)
$NbN_{0.87}$	62.4	62.4	62.4	(g)
	60.5	60.8	60.8	(g)
$NbN_{0.80}$	48.2	46.85	46.6	(g)
	48.15	46.55	46.2	(g)
$NbN_{0.97}$	85			(c)
$NbN_{0.75}$	90			(c)
TaN (hex)	128			(c)

and shape of the pores are generally unknown. Corrections for porosity use empirical formulas of the type

$$\rho \text{ (actual)} = \rho \text{ (measured)} \times P^{3.5}$$

where P is the ratio of the actual density to the X-ray density (11). It would appear obvious that measurements on single crystals with low residual porosity would yield the most accurate results. Yet Piper (12) has found that a single crystal of $TiC_{0.94}$ has a resistivity value substantially higher than that of the same material which was crushed and hot-pressed to 85% theoretical density (178 $\mu\Omega$-cm versus 90 $\mu\Omega$-cm). The reason for this difference is not clear, although the subsequent heating may have altered the metal-to-nonmetal ratio or the impurity level.

Rudy and Benesovsky (2), using hot-pressed very dense samples, measured electrical resistivities of several pseudobinary solid solutions. In most of these pseudobinaries, the variation of ρ with composition exhibits a small maximum, as might be expected in analogy to the behavior in binary alloys (see Fig. 2). Pseudobinary solutions involving VC as one component exhibit an anomolous resistivity behavior (see Fig. 3). Similar anomalies involving pseudobinaries with VC or VN have been observed in the superconducting properties (see Chapter 7).

Electrical resistivities of carbides and nitrides are characterized by an

---

[a] See also Fig. 1.
[b] Studied as a function of temperature from 0 to 1600°C.
[c] Stoichiometric VC does not exist; these phases have a carbon-to-vanadium ratio not exceeding 0.88, the phase boundary.
[d] Measured on samples with considerable porosity. The porosity was corrected with the formula ρ (actual) $= \rho$ (measured) $\times P^{3.5}$.

References for Table I

a. E. Rudy and F. Benesovsky, *Planseeber. Pulvermet.* **8**, 72 (1960).
b. P. Costa and R. R. Conte *in* "Compounds of Interest in Nuclear Reactor Technology" (J. T. Waber, P. Chiotti, and W. N. Miner, eds.), Inst. Metals Div., Spec. Rep. No. 13, p. 3. Edwards, Ann Arbor, Michigan, 1964.
c. G. V. Samsonov, "High-Temperature Materials, Properties Index." Plenum Press, New York, 1964.
d. F. W. Glaser and W. Ivanick, *J. Metals* **4**, 387 (1952).
e. F. W. Glaser and D. Moskowitz, *Powder Met. Bull.* **6**, 178 (1953).
f. A. Munster and K. Sagel, *Z. Phys.* **144**, 139 (1956); *Angew Chem.* **69**, 281 (1957).
g. N. Pessall, J. K. Hulm, and M. S. Walker, Westinghouse, Research Laboratories, Final Rep. AF 33(615)-2729 (1967).
h. H. J. Fink, A. C. Thorsen, E. Parker, V. F. Zackay, and L. Toth, *Phys Rev. A* **138**, 1170 (1965).
i. L. E. Toth and J. Zbasnik, *Acta Met.* **16**, 1177 (1968).

unusually small temperature coefficient (11–15). Resistivities of pure elements decrease by a factor of about 100 between room temperature and
liquid helium temperatures. Piper (12) finds that ρ for TiC decreases only
10% in this temperature range. Pessall et al. (11) find the resistivity for
nearly stoichiometric NbN increases slightly with decreasing temperature.

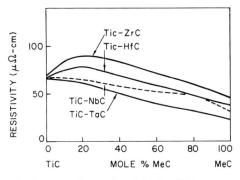

Fig. 2. The variation of the electrical resistivity in several pseudobinary carbide systems shows a small
maximum typical of the behavior in
transition-metal binary systems. [After
Rudy and Benesovsky (2).]

The increase is small with $(\Delta\rho/\Delta T)_{av} = -3.5 \times 10^{-3}$ $\mu\Omega$-cm/°K in the range
4–300°K. Most carbides and nitrides have a $\Delta\rho/\Delta T = +5 \times 10^{-2}$ $\mu\Omega$-cm/°K.

Thin nitride films of Nb–Ti–N and Nb–Zr–N solid solutions have a very
small temperature dependence of the resistivity (15). Films, 1000–2000 Å
thick, of Nb–Ti–N deposited on pyrex have a temperature coefficient less

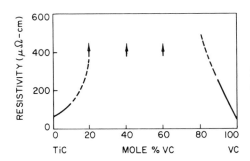

Fig. 3. A pseudobinary solution
between VC and TiC exhibits an
anomalous resistivity behavior. [After
Rudy and Benesovsky (2).]

than 30 ppm/°C in the temperature range 12–400°K. This material could be
useful as temperature-independent thin-film resistors. Nb–Zr–N thin films
have a resistivity which increases by about 10% on cooling from room
temperature to helium temperatures.

The origin of the high residual resistivities at low temperatures is probably
associated with high vacancy concentrations on carbon or nitrogen sites.
Vacancies on the metal sublattices also contribute to residual resistivity.
Pessall et al. (11) have postulated that high residual resistivity and its negative
temperature coefficient in NbN are associated with a high degree of disorder

in both metal and nonmetal vacancies. They compare the resistivity behavior of NbN to TiO, in which the coefficient is also negative and both metal and nonmetal vacancies exist.

III. Hall Coefficients

Hall coefficients yield information about the number of electron carriers in the conduction band, the band shape as a function of electron concentration, and, when measured in conjunction with the resistivity, the mobility of the carriers. Table II lists some of the more recent results measured on either single crystals or very dense carbides and nitrides. The data show considerable disagreement, possibly the result of different deviations from stoichiometry or impurity levels. Unfortunately, these parameters were not always specified. Other data on generally less dense samples may be found elsewhere (1).

The Hall coefficient R_0 is sensitive to the carbon content in TaC_{1-x} (Fig. 4) (7). This system, however, has been the only one studied over a wide

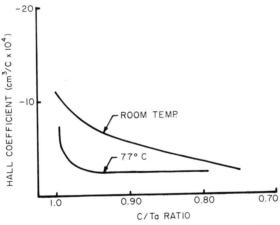

Fig. 4. The Hall coefficient increases sharply with carbon concentration at compositions close to the stoichiometric in TaC_{1-x}. [After Santoro and Dolloff (7).]

composition range. Over a narrower range, R_0 does not vary appreciably in TiC_{1-x} at room temperature (8).

A number of investigators have calculated the carrier concentrations by assuming a simple one-band model equation,

$$R_0 = M/enN_0\rho_0,$$

where n is the carrier concentration, M the formula weight, and ρ_0 the X-ray density (16–18). All values of R_0 included in Table II and Fig. 4 are negative; thus the dominant carriers are electrons. Investigators conclude that the number of electrons per formula increases from nearly zero for the fourth-group carbides to nearly one for the fifth-group carbides. The fourth-group nitrides also have a carrier concentration of about one per formula, and this

TABLE II

HALL COEFFICIENTS OF CARBIDES AND NITRIDES

Phase	Hall coefficient, $R_0(cm^3/C) \times 10^4$			Hall mobility, cm²/ volt·sec (room temp)	Conduction electrons/ formula (room temp)	Ref.
	Room temp value	77°K	0°K Extra- polated			
$TiC_{0.96}$	−7.04			11	0.18	(a)
TiC	−(14.1–13.7)	−(24.8–27.6)			0.08	(b)
$TiC_{0.928}$	−15.5	−22.4		8.96		(c)
$TiC_{0.939}$	−15.0	−26.4		9.20		(c)
$TiC_{0.969}$	−16.6	−34.2		12.21		(c)
$TiC_{>0.969}$	−15.0	−31.7		15.80		(c)
TiC	−6.7			-		(d)
ZrC	−19.3			28.6	0.09	(a)
$ZrC_{0.96}$			−30.1			(e)
ZrC	−9.4					(d)
HfC	−18.1			28.9	0.09	(a)
HfC	−12.4					(d)
$VC_{0.7}$	−0.91			0.99	1.2	(a)
NbC			−1.55		0.90	(e)
NbC	−1.3					(d)
$NbC_{0.97}$	−1.28			3.2	1.04	(a)
TaC	−0.64			0.71	2.2	(a)
TaC	−1.1					(d)
TaC			−1.30		1.07	(e)
$ZrN_{0.95}$			−1.75		0.87	(e)
NbN			−0.39			(e)

References for Table II

a. T. Tsuchida, Y. Nakamura, M. Mekata, J. Sakurai, and H. Takaki, *J. Phys. Soc. Jap.* **16**, 2453 (1961).
b. J. Piper, *J. Appl. Phys.* **33**, 2394 (1962).
c. W. S. Williams, *Phys. Rev. A* **135**, 505 (1964).
d. G. V. Samsonov and V. N. Paderno, *Planseeber. Pulvermet.* **12**, 91 (1964).
e. J. Piper in "Compounds of Interest in Nuclear Reactor Technology" (J. T. Waber, P. Chiotti, and W. N. Miner, eds.), Inst. Metals Div., Spec. Rep. No. 13 p 29. Edwards, Ann Arbor, Michigan, 1964.

number increases in solid solutions with NbN. Figure 5 shows the number of carriers as a function of composition in several pseudobinary systems.

The temperature dependence of R_0 has been determined for many compositions. Nearly stoichiometric fourth-group carbides are temperature-dependent, although there is disagreement on this point [compare Piper's

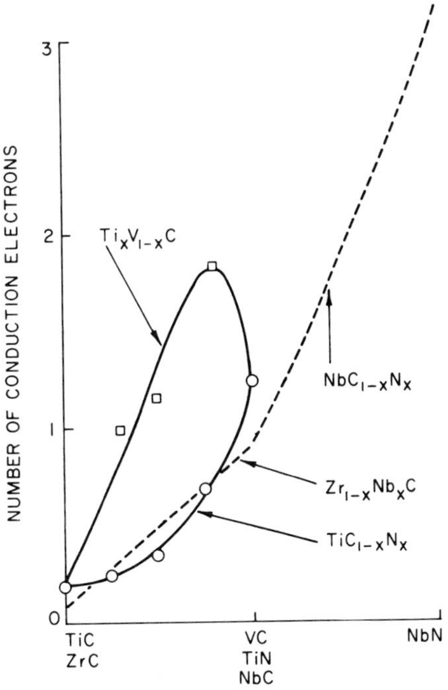

Fig. 5. The number of electron carriers increases as the composition changes from the fourth-group carbides to higher groups. - - - - : data of Piper (*17*) ; □, ○ : data of Itoh *et al.*(*18*).

(*17*) and Tsuchida's (*16*) data for TiC, Fig. 6]. The fifth-group carbides and fourth and fifth-group nitrides are relatively temperature-independent.

From the temperature dependence of R_0 and the variation of n with composition, several investigators (*16–18*) have concluded that a new band is beginning to be filled by electrons at the fourth-group carbides. As the composition changes to higher groups, the number of electrons in this band increases (Fig. 5). They further suggest that the band structure of fourth- and fifth-group carbides and nitrides has the same rigid shape. To account for variation in properties with composition, the height of the Fermi level is adjusted according to the valence electron concentration (VEC), i.e., the number of valence electrons per formula—9 for ZrN, 9 for NbC, etc. Inspection of Table II and Fig. 5 shows that the correlation is approximately valid for nearly stoichiometric carbides and nitrides. We shall see that several other electrical and magnetic properties can be correlated in the same manner.

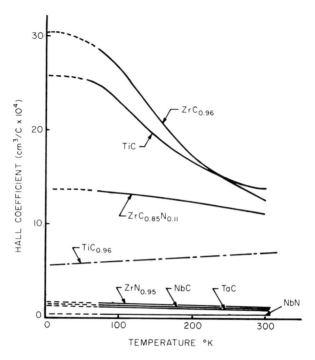

Fig. 6. The temperature dependence of R_0 for several carbides and nitrides. ——:
data of Piper (*17*); — — —: data of Tsuchida *et al.* (*16*).

Santoro and Dolloff (*7*) have shown, however, that the correlation does
not hold for nonstoichiometric TaC_{1-x}. According to the VEC correlation,
R_0 values should increase as the VEC decreases from 9 to 8, yet the opposite
behavior is observed (Fig. 4). Their values of R_0 for nearly stoichiometric
TaC are much higher than the values listed in Table II.

IV. Magnetic Susceptibilities

Table III lists the magnetic susceptibilities of several nearly stoichiometric
monocarbides and nitrides with the *B*1 crystal structure. This susceptibility
is strongly dependent on the nonmetal-to-metal ratio which may explain
some of the differences in values listed in Table III for supposedly the same
phase. Figure 7 illustrates susceptibility variations with metal-to-nonmetal
ratios for several binary systems; variation is particularly pronounced for
the fourth-group carbides and VC_{1-x} (*10, 19*).

TABLE III

MAGNETIC SUSCEPTIBILITIES FOR CARBIDES AND NITRIDES

Alloy system	$X_g \times 10^6$ emu/mole	Ref.
TiC[a]	-7.5	(a)
TiC	$+5.7$	(b)
ZrC[a]	-30	(a)
ZrC	-23	(b)
HfC[a]	-37	(a)
HfC	-25.5	(b)
$VC_{0.85}$	$+35$	(a)
$VC_{0.84}$	$+28$	(c)
NbC[a]	$+20$	(a)
NbC	$+15.3$	(b)
$NbC_{0.98}$	$+15$	(d)
TaC[a]	$+12$	(a)
TaC	$+9.3$	(b)
TaC	$+20$	(c)
$TaC_{0.97}$	$+12$	(e)
TiN	$+38$	(c)
ZrN	$+22$	(a)
VN	$+130$[b]	(c)
$NbN_{0.91}$	$+31$	(d)

[a] Values extrapolated to the stoichiometric composition from the known variation of the susceptibility with X/Me ratio.

[b] Doubtful value—given by a Hondo extrapolation $(1/H = 0)$ on a specimen showing ferromagnetic impurities.

References for Table III

a. H. Bittner and H. Goretzki, *Monatsh. Chem.* **93**, 1000 (1962).
b. G. V. Samsonov and V. N. Paderno, *Planseeber. Pulvermet.* **12**, 19 (1964).
c. P. Costa, PhD, Thesis, Orsay, 1964, reported by P. Costa, *in* "Anisotropy in Single-Crystal Refractory Compounds" (F. W. Vahldiek and S. A. Mersol, eds.), Vol. I, p. 151. Plenum Press, New York, 1968.
d. T. H. Geballe, B. T. Matthias, J. P. Remeika, A. M. Clogston, V. B. Compton, J. P. Maita, and H. J. Williams, *Physics (Long Island City, N.Y.)* **2**, 293 (1966).
e. L. B. Dubrovskaya and I. I. Matveyenko, *Phys. Metals Metallogr. (USSR)* **19**, No. 2, 42 (1965).

Bittner and Goretzki (*20, 21*) also measured the susceptibilities of several nearly stoichiometric pseudobinary systems between monocarbides and mononitrides. They conclude from these measurements that magnetic susceptibility can be correlated with valence electron concentration (VEC) and with the concept that the band structure of fourth- and fifth-group carbides and nitrides has nearly the same rigid shape. (See the previous

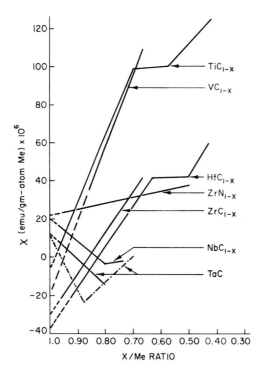

Fig. 7. Variation of the magnetic susceptibility with nonmetal-to-metal ratios for several monocarbides and nitrides. ——: data of Bittner and Goretzki (*10*); ———: data of Dubroskaya and Matveyenko (*19*).

section on Hall coefficients for an introduction to the VEC concept; the VEC theory of bonding will be discussed in Chapter 8.)

Figure 8 shows the magnetic susceptibility for several carbides and nitrides plotted against VEC. A fair correlation exists. A minimum in the density of states is indicated at the fourth-group carbides. The density rapidly increases at VEC compositions less than 8. The VEC correlation for susceptibilities also appears valid for the nonstoichiometric carbides NbC_{1-x} and TaC_{1-x}. Figure 8 clearly shows, however, that the VEC correlation does not apply to ZrN_{1-x} and VC_{1-x}. Hall constant and γ measurements also show that the VEC parameter cannot be used to correlate the properties as a function of deviations from stoichiometry.

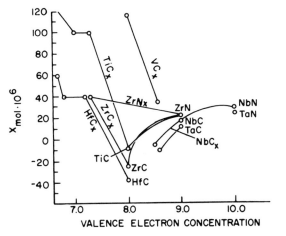

Fig. 8. Room temperature magnetic susceptibilities plotted as a function of valence electron concentration. [Data from Bittner and Goretzki (*10, 20, 21*).]

Costa and Conte (*3*) have correlated the susceptibilities by assuming that C donates 1.5 electrons and N donates 2 electrons to the *d*-like bands of the transition element. Figure 9 shows this correlation. All values are grouped around one mean curve, except for the VC_{1-x} data.

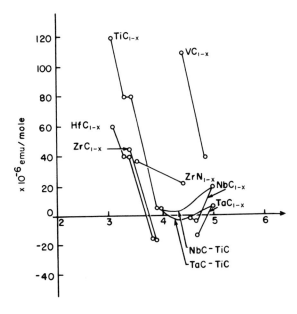

Fig. 9. Magnetic susceptibilities plotted as a function of assumed number of *d*-like electrons. Carbon is assumed to donate 1.5 electrons to the *d* bands and nitrogen, 2 electrons. [After Costa and Conte (*3*).]

V. Electronic Specific Heat Coefficient

Electronic specific heat coefficients γ of carbides and nitrides have already been tabulated in Table I of Chapter 4. γ values have been studied as a function of composition, defect structure, and crystal structure. Since γ is proportional to the density of electron states at the Fermi level, these studies, in conjunction with the magnetic susceptibility, Hall and other measurements, help interpret the electronic structure of carbides and nitrides. They are particularly useful in evaluating the basis for the assumption that band structures of the refractory carbides and nitrides all have approximately the same rigid shape and that the variation of properties can be explained by adjusting the height of the Fermi level according to the VEC or some other parameter.

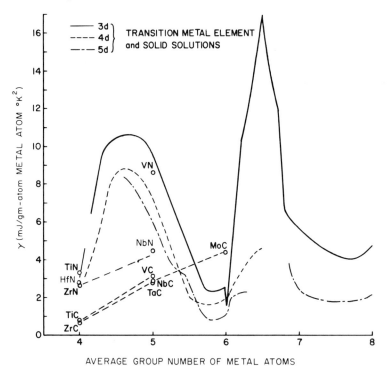

Fig. 10. γ values for carbides and nitrides are generally lower than those for transition-metal elements and solid solutions. The γ values for the carbides and nitrides were taken from Table I of Chapter 4. The curves for the transition metal elements and solid solutions are smoothed curves drawn through many individual γ values for a particular composition. The γ values are from a compilation by Heiniger *et al.* (22).

Figure 10 illustrates an unusual feature about the γ values of carbides and nitrides (22). Compared to the values of most transition-metal elements and their solid solutions, the carbide and nitride values are very low, indicating low density-of-states values at the Fermi level. Several investigators (23, 24) have suggested that the two high density-of-states d-like subbands in the transition-metal elements band structure are split far apart in energy in forming carbides and nitrides and that a low density-of-states conduction band exists between the two subbands. This interpretation is consistent with Hall and susceptibility measurements. The Hall measurements indicate that a new band is forming at the stoichiometric fourth-group carbides. The fourth-group carbides also have the lowest γ values and susceptibility values. Decreasing the electron concentration by removing carbon in the fourth-group monocarbides increases markedly the susceptibility values. This increase indicates that the Fermi level is now in the high density-of-states d-like subband. Between the stoichiometric fourth-group monocarbides and the sixth-group monocarbides, the relatively low γ values indicate that the Fermi level is in a broad conduction band. The number of electrons in this band, as deduced from the Hall measurements, again substantiates this interpretation.

Table IV illustrates another unusual feature about the γ values of carbides and nitrides. For a given group number, nearly stoichiometric monocar-

TABLE IV

γ VALUES OF MONOCARBIDES BY GROUP NUMBER[a]

IV		V		VI	
TiC	0.51–0.75	VC	3.15	—	
ZrC	0.75	NbC	2.83	MoC	4.40
HfC	0.75	TaC	3.2	—	

[a] Values (mJ/gm-atom Me), refer to compositions closest to stoichiometry. (Consult Table I, Chapter IV for references.)

bides and nitrides have very nearly identical γ values. As Table V shows, the γ values of nearly stoichiometric fourth-group nitrides are also very similar to those of fifth-group carbides, and the same similarity holds between NbN and MoC. These relationships suggest that γ values can be correlated with the valence electron concentration, VEC, just as the magnetic susceptibilities, Hall constants, and superconducting properties of nearly stoichiometric carbides and nitrides can be correlated.

TABLE V

COMPARISON OF γ VALUES OF NEARLY
STOICHIOMETRIC CARBIDES AND
NITRIDES BY GROUP NUMBER [a]

Group IV nitride		Group V carbide	
TiN	2.5–3.3	VC	3.15
ZrN	2.67	NbC	2.83
HfN	2.73	TaC	3.2
Group V nitride		Group VI carbide	
NbN	4.08–4.56	MoC	4.4

[a] Values are in millijoules per gram-atom
Me.

When nonstoichiometric carbides and nitrides are studied, the simple VEC correlation fails. Figure 11 shows the variation of γ with nonmetal-to-metal ratio for several carbides and nitrides. For these carbides, γ decreases monotonically with decreasing carbon content from the nearly stoichiometric monocarbide to the subcarbide. For NbN_{1-x}, γ increases or remains nearly

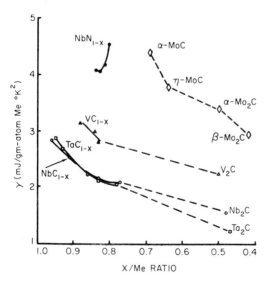

Fig. 11. The variation of γ with nonmetal-to-metal ratio shows a different behavior for nitrides and carbides. Data are from Table I of Chapter 4.

the same as the nitrogen-to-metal ratio decreases. This dissimilar behavior shows that nonstoichiometric carbides and nitrides cannot be regarded as isoelectronic phases, even though the simple VEC empirical correlation works well for the stoichiometric phases.

Several investigators have proposed that the variation of γ with electron concentration in carbides and nitrides should be similar to that found in transition-metal alloys, allowing for an electron transfer from the nonmetal to the d bands of the transition elements (*3*, *25*, *26*). As Fig. 10 shows, however, this interpretation does not appear valid, even if a relative shift in the two curves due to an electron transfer is assumed; the γ values for carbides and nitrides are generally lower than transition-metal elements. On the other hand, Fig. 12 shows that γ values for nonstoichiometric monocarbides

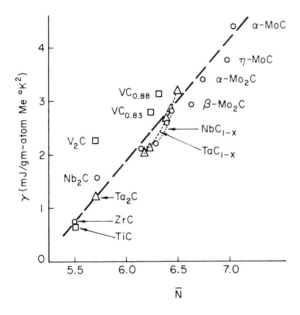

Fig. 12. The γ values of carbides cluster about a single curve when plotted against $\bar{N}$ defined as the group number of transition element plus $1\frac{1}{2}$ times the carbon concentration. γ values were taken from Table I of Chapter 4.

and subcarbides can be correlated with the group number of the transition metal plus 1.5 times the carbon concentration per formula. All the γ values cluster about a single line. This parameter is equivalent to Costa and Conte's assumed number of d-like electrons (*3*). This correlation is striking because it describes both the nonstoichiometric monocarbides with the NaCl structure and the subcarbides with hexagonal or orthorhombic crystal structures. Since there is little correspondence between the γ values of these carbides and those of the transition elements with the same number of d-like electrons,

TABLE VI

SUMMARY OF LOW-TEMPERATURE HEAT CAPACITIES OF MOLYBDENUM CARBIDES WITH CLOSELY RELATED STRUCTURES

Phase	Composition	Structure and sequential ordering of close-packed Mo layers	Lattice parameter (Å)	γ (mJ/gm atom-Mo °K²)	θ_D (°K)
α-MoC$_{1-x}$	MoC$_{0.69}$	$B1$ (NaCl) —ABCABC—	$a = 4.281$	6.40	610
η-MoC$_{1-x}$	MoC$_{0.64}$	Hexagonal —ABCBABCBA—	$a = 3.010$ $c = 14.62$	3.79	536
α-Mo$_2$C	MoC$_{0.5}$	Orthorhombic —ABAB— (carbon ordered)	$a = 4.736$ $b = 6.024$ $c = 5.217$	3.41	473
β-Mo$_2$C	MoC$_{0.42}$	Hexagonal L_3' —ABAB— (carbon disordered)	$a = 2.995$ $c = 4.730$	2.94	492
Mo$_2$BC	Mo$_2$BC[a]	Complex Orthorhombic	$a = 3.086$ $b = 17.35$ $c = 3.047$	4.25	529
Mo$_3$Al$_2$C	Mo$_3$Al$_2$C[a]	β-Mn type (complex cubic)	$a = 6.865$	9.30	248

[a] Nominal composition.

the physical significance of the correlation is unclear. The correlation could be interpreted as indicating an electron transfer from carbon to the metal atoms. A similar correlation could not be found for the nonstoichiometric nitrides.

Several investigators (24, 26–28) have noted the close relationship of the γ values of the subcarbides and those of the nonstoichiometric monocarbides. These results suggest that the shape of the bands is rather insensitive to the crystal structure (24, 26). More extensive studies (24, 27) have been made on the molybdenum carbides. Results show that γ values are relatively insensitive to crystal structure changes, provided the crystal structures are closely related (see Table VI). In binary molybdenum carbides, the metal atoms occupy a fcc or hcp sublattice and the carbon atoms occupy a fraction of the interstitial sites in an ordered or random fashion; first nearest neighbors are nearly identical in all structures. Crystal structures of Mo_2BC and Mo_3Al_2C are related to those of the binary phases by a common Mo_6C octahedral coordination polyhedron. With the exception of Mo_3Al_2C, the γ values for all the phases in Table VI are very similar.

VI. Thermoelectric Power

Values of the thermoelectric power S are tabulated in Table VII. The values are negative and decrease in absolute magnitude from the fourth to fifth group carbides and again from the fourth- to fifth-group nitrides. Lye (29) and Costa and Conte (3) studied the variation of S in TiC_{1-x} as a function of carbon concentration. As carbon is removed from the lattice in TiC_{1-x}, S becomes less negative or even slightly positive.

Interpreting the thermoelectric power in carbides is complicated because a number of assumptions must be made about the Fermi energy, the electron mean free path, and the band shape. Lye (29) attempted a detailed analysis of the effect in TiC_{1-x}. He found that the results could be explained if he assumed that the Fermi energy increased as the carbon content increased in TiC_{1-x}. The results then indicate that there is a rapid decrease in the density of d states as the carbon content is increased. This result agrees with the interpretation of the magnetic susceptibility data in TiC_{1-x}. In order for the Fermi energy to increase as carbon is added to the lattice, it is necessary for carbon to act as an electron donor to the d bands.

TABLE VII

THERMOELECTRIC POWER OF CARBIDES AND NITRIDES

Phase	S (μV/$^{\circ}$K) (room temp)	Reference
TiC	-11.2	(a)
	-10.81	(b)
	-15.5	(c)
	-12.6	(d)
$TiC_{0.6}$	-4.0	(c)
$TiC_{0.5}$	$+1.0$	(c)
TiC_{1-x}	-8 to -1	(e)
ZrC	-11.3	(a)
	-12.7	(d)
HfC	-12.5	(f)
	-12.5	(d)
VC	-4.12	(b)
NbC	-4.0	(a)
	-5.4	(d)
TaC	-5.0	(a)
	-1.5	(c)
	-6.6	(d)
TiN	-7.8	(g)
ZrN	-4.8	(g)
HfN	-3.0	(g)
VN	-4.6	(g)
NbN	-1.65	(g)
TaN	-1.6	(g)

References for Table VII

a. S. N. L'vov, V. P. Nemtchenko, and G. V. Samsonov, *Dokl. Adad. Nauk SSSR* **135**, No. 3, p. 577 (1960) [Transl. *Sov. Phys.-Doklady* **5**, 1334 (1961).]
b. S. Noguchi and T. Sato, *J. Phys. Soc. Jap.* **15**, 2359 (1960).
c. P. Costa and R. R. Conte *in* "Compounds of Interest in Nuclear Reactor Technology" (J. T. Waber, P. Chiotti, and W. N. Miner, eds.), Inst. Metals Div. Spec. Rept. No. 13, p. 3. Edwards, Ann Arbor, Michigan, 1964.
d. G. V. Samsonov and V. N. Paderno, *Planseeber. Pulvermet.* **12**, 19 (1964).
e. R. G. Lye *J. Phys. Chem. Solids.* **26**, 407 (1965).
f. S. N. L'vov, V. P. Nemtchenko, and G. V. Samsonov, *Porosh Met.* **4**, No. 10, 3 (1962).
g. G. V. Samsonov and T. S. Verkhoglyadova, *Dokl. Akad. Nauk SSSR* **142**, No. 3, 608 (1962).

VII. Thermal Conductivities

Like many other properties of carbides, their thermal conductivities are unusual. Radosevich and Williams (30, 31) conducted several very interesting experiments on the low-temperature behavior and Taylor (32, 33) on the high-temperature behavior. In both regions, the properties of the carbides differ significantly from those of a typical transition metal. For a general discussion of thermal conductivities of metals the review article of Mendelssohn and Rosenberg (34) and the model of Callaway (35) should be consulted.

In a metal, the total heat transport is the sum of the heat conducted by electrons and that conducted by the lattice; that is, the total heat conductivity K is the sum of the conductivities of electrons and phonons or

$$K = K_e + K_g$$

where K_e is the conductivity due to electrons and K_g is the conductivity due to phonons. It is possible to separate at all the individual scattering terms from one measurement of $K(T)$.

In the case of carbides, Radosevich and Williams found that the contribution to K from electrons is small and never exceeds 10% of the total conductivity; thus, the conductivity is principally by phonons and a T^2 dependence is expected for phonon–electron scattering. Instead, the data indicate that at low temperature (1–10°K) the phonon contribution varies as $T^{1.3}$. The explanation cannot be point defect scattering since the conductivity increases with increasing deviations from stoichiometry (see Fig. 13).

To explain these results, Radosevich and Williams suggest that the electron–phonon interaction is strong but that the adiabatic approximation in not valid in carbides. This approximation states that the electron wave function adjusts continuously to the ion motion and its use in the ordinary theory of conductivity yields a T^2 dependence for phonon–electron scattering. The condition to use the adiabatic approximation is

$$q\lambda_e > 1$$

or the wavelength of phonons is shorter than the mean free path of the scattered electron. This approximation breaks down because the electron mean free path in carbides is small (1–50 Å depending on composition and whether the specimen is in bulk or film form) and at low temperatures the principal phonons have long wavelengths. They propose that the principal mechanism responsible for the low lattice conductivity is phonon scattering by electrons and show that the $T^{1.3}$ dependence could arise from a reformulation of the theory without the adiabatic approximation.

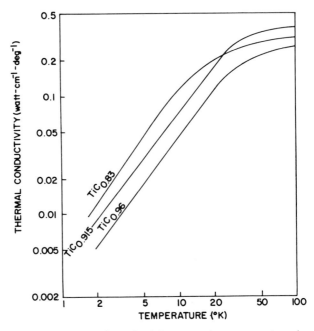

Fig. 13. In TiC_{1-x} the thermal conductivity at very low temperatures increases with increasing deviations from stoichiometry. If point defects controlled the scattering process, the opposite behavior would be expected. [After Radosevich and Williams (*30*).]

To test this hypothesis, the thermal conductivity of $NbC_{0.96}$ in the superconducting state was compared to that in the normal state. When the heat transfer is principally by electrons, the onset of superconductivity results in a decreased thermal conductivity since the superconducting electrons are coupled in pairs. This decrease is the case for most superconductors. For $NbC_{0.96}$, the lattice thermal conductivity increases greatly below T_c due to decreased phonon-electron scattering. This result supports the hypothesis that phonons are principally scattered by conduction electrons in carbides, even though the T^2 dependence is not observed (see Fig. 14). Additional evidence for an unusually strong electron–phonon interaction in carbides is given in the next chapter on superconductivity.

At high temperatures the thermal conductivity of carbides is also unusual as Taylor (*32, 33*) discovered. Above 500°C the conductivity in TiC is nearly equally divided between the phonon and electron contributions. The electron contribution is estimated by the Wiedermann–Franz law, $K_e = L_0 T / \rho$ on the phonon contribution by the difference $K_g = K - K_e$. Accord-

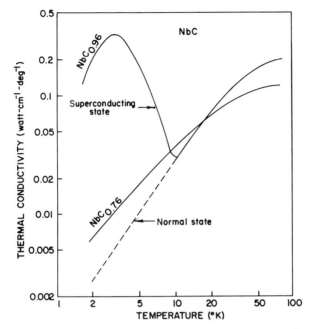

Fig. 14. In superconducting $NbC_{0.96}$ the thermal conductivity is significantly greater than in the normal state. The thermal resistivity is due to phonon scattering by electrons, and in the superconducting state these electrons can no longer scatter phonons. Hence, the conductivity increases. [After Radosevich and Williams (*31*).]

ing to theory (*34*), the lattice contribution should decrease as $1/T$, since the conductivity is limited by umklapp processes. Instead, K_g was found to be nearly independent of T. Williams (*36*) and Radosevich and Williams (*37*) have analyzed this behavior and have shown that it is due to strong phonon–point defect interactions. These interactions lessen the temperature dependence of K_g. The phonon scattering is due primarily to force constant changes rather than mass changes which are introduced by the vacancies.

VIII. Soft X-Ray Emission

Experimental information about the electron density of states, the widths of various bands and the relative energy spacing between different bands can be found from soft X-ray emission and absorption studies. The spectra intensity cannot be exactly related to the density of states nor is an unambiguous assignment of bands possible. Nevertheless, valuable information about the band structure can be obtained.

Fischer and Baun (*38*) carefully studied the spectra for Ti, TiC, TiN, and several titanium oxides and their results give insight into the band structures of TiN and TiC. Figure 15 shows the spectra for Ti, TiC, TiN, and TiO.

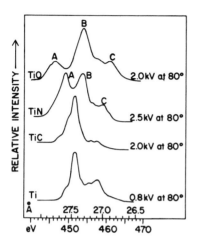

Fig. 15. A soft X-ray emission spectrum for Ti. $L_{II, III}$ in TiC is similar to that for pure Ti while that for TiN is more similar to the one for TiO. Since the bonding in TiO has a strong ionic component with oxygen negatively charged, the spectra indicate that nitrogen is also negatively charged in TiN. In TiC, however, the charge transfer may be in the opposite direction. [After Fischer and Baun (*38*).]

The spectrum for TiC most closely resembles that of Ti while that for TiN resembles the one for TiO. Since TiO is known to have a large ionic component to the bonding with electrons being transferred from Ti to the oxygen bands, they interpret the spectra as indicating that TiN also has a strong ionic component in the same sense. They interpret the TiC spectrum as indicating a transfer of electrons from C to Ti. This spectrum for TiC was found to agree with the band structure calculation of Lye and Logothetis (*40*) (discussed in Chapter 8) in which an electron transfer of about 1.3 electrons/carbon atom was assumed.

It is important to emphasize that the results of Fischer and Baun clearly show that the band structures for TiC and TiN differ considerably. As discussed in the next section, most correlations between electron concentration and properties have incorrectly assumed a rigid band model approximately applicable for most carbides and nitrides.

Holliday (*41, 42*) has compared the soft X-ray emission spectra for the *K* emission band in a series of transition metal carbides with those spectra for diamond and graphite. He observed that the shape of the carbon *K* emission band for a given group of carbides was nearly the same. The shape of the carbon *K* bands for group IV carbides is a single, nearly symmetrical peak, while for group V the bands have a narrow main peak with a hump on the higher side. The shape of the band is rather insensitive to the crystal structure; the shape of the band for TaC is nearly the same as that for Ta$_2$C. This result would also indicate that the band shape should not be sensitive to the C/Me ratio. For group VI and higher the shapes of the bands in the carbides more closely resemble that for graphite than they do for the group IV carbides. The carbon K bands are broader than they are in the fourth or fifth group carbides. The increased width of the bands corresponds to a decreased thermodynamic stability of the carbides.

Holliday explains his results in the following manner: In the group IV carbides the carbon band would have only one σ-electron and no π-electrons. In the group V carbides the carbon bands would have one and a fraction σ-electrons and one π-electron, and in the group VI carbides the bands would have the same distribution as in the band for graphite (one π- and three σ-electrons). Thus in the case of the group IV and V carbides the electron transfer is from carbon to the d bands of the transition metal.

While the actual number of transferred electrons cannot be accurately assessed, an important result of Holliday's experiments is the evidence that the number of transferred electrons varies with the group number of the carbide. Previous attempts to explain the physical and bonding properties of the carbides have not adequately introduced this concept. It seems plausible, however, that the number does vary with group number.

In the group VI elements, the heats of sublimation and the shape of the band structure indicate that the bonding portion of the band is just filled. One of the results of a transfer of electrons from carbon to the metal d bands in group IV and V carbides is an increased strength of the metal–metal bonds. An electron transfer in the group VI carbides does not have the same effect but tends to weaken the bond strength because the antibonding portion of the bands is being filled. A similar idea has been proposed for the transition metal borides, for which on the basis of low temperature heat capacity measurements discrepancies in the electron transfer theory can only be explained by assuming a variable transfer of electrons depending upon the group number (43). In the earlier groups the transfer from boron to the d bands of the metal is greatest, while in the later groups only a small transfer occurs. In speaking of an electron transfer for the carbides, one has to remember that the $2p$ bands of carbon and the $3d$ band of the first row transition metals are probably strongly hybridized so that much of the individual character of the wave functions is lost.

Nemnonov and co-workers (44–47) have investigated the soft X-ray spectra of several titanium compounds. They find that their results are consistent with an electron transfer to nitrogen in TiN and a transfer from carbon in TiC. They compare the soft X-ray emission spectra for titanium in TiC, TiN, and TiO with the theoretical band structure calculations of Ern and Switendick (48) and they find very good quantitative agreement with the energy placement of the bands and the shape of the bands.

As pointed out earlier, Fischer and Baun assert that their results are consistent with the band model of Lye and Logothetis, which assumes an electron transfer of about 1.3 electrons from carbon to the metal d-bands. Nemnonov finds agreement with the Ern and Switendick model, in which the $2p$ and $3d$ bands are highly hybridized but in which the placement of the $2p$ band is lower in energy than in the Lye and Logothetis (40) model. Thus, in the Ern and Switendick (48) model, electron transfer from carbon is small and, in fact, may occur in the opposite direction. The point which should be emphasized here is that an unambiguous decision about the degree and direction of the electron transfer in the carbides cannot be made on the basis of soft X-ray emission studies alone. Two band models with different starting assumptions have been used to interpret the X-ray results. Both seem to give equally good agreement with experiment.

IX. Limitations of Electron Correlations, Fact or Fiction

The literature on carbides and nitrides contains numerous attempts to correlate electrical and magnetic properties with electron concentrations (3, 25, 49–52). It is tempting to speculate about why so many correlations have been proposed. First, many correlations have had reasonable success in interpreting one or more properties. Second, before the availability of sophisticated band structure calculations, the correlations were a means, although a poor one, of interpreting bonding in these compounds. Third, since much data on carbides and nitrides are still not available, the correlations, when properly applied and not over-extended, are useful in predicting property values.

The two most popular correlations are the valence electron concentration (VEC) due to Bliz (52) and the electron transfer theory (3, 25, 49–51).

The VEC for the formula MeC_x or MeN_x is the sum of the group number of Me and X times the group number of C or N. In the VEC correlation the band structure is assumed to be rigid and applicable to both carbides and nitrides. Properties are determined for different compositions by adjusting the height of the Fermi level according to the VEC.

In the electron transfer theory as proposed by Dempsey (25) the electron band structure for the interstitial compounds closely resembles that for the transition elements. Carbon donates one electron to the d band and nitrogen two electrons to shift the position of the Fermi level with respect to its position in the transition metals. Thus the properties of NbC should resemble those of Mo in this scheme. There are several variants to the correlation (3, 49–51).

We have throughout this chapter referred to these correlations. Sufficient experimental data now exist to evaluate critically these correlations and discuss their limitations. Additional reference to the correlations will be given in the next chapter on superconductivity and their theoretical basis will be discussed in Chapter 8 on bonding.

PREMISE 1. *The carbides and nitrides are isoelectronic.*

This premise is basic to both the Bilz and Dempsey correlations. The idea that nearly all carbides and nitrides can be described by a rigid band model with the Fermi level dependent only upon the electron concentration or transfer has its foundation in the behavior of stoichiometric carbides and nitrides. As illustrated in Table V, γ values for stoichiometric group V carbides and group IV nitrides are nearly equal, as are those for group VI carbides and group V nitrides. These results explain the success of the isoelectronic concept in correlating the properties dependent upon the density of states at the Fermi level for stoichiometric carbides, nitrides and carbide–nitride solid solutions. When, however, the properties of *nonstoichiometric* carbides and nitrides are considered, the premise fails. While the data for nonstoichiometric nitrides is scarce, their properties vary in a nearly opposite manner to those of nonstoichiometric carbides. Furthermore, soft X-ray measurements show that the band structures for carbides and nitrides differ considerably. Thus, the general conclusion is that this premise is not valid.

PREMISE 2. *The band structure for carbides and nitrides is closely related to that for transition metals.*

This premise is part of the Dempsey transfer theory and several others. It is useful in conjunction with Dempsey's proposed direction of electron transfer in correlating melting points, resistivities and other properties with those of transition metals. Furthermore, many of the carbide and nitride properties are similar in magnitude to those of the transition metals. Low temperature heat capacities and soft X-ray spectra show, however, that this premise is a poor approximation. A more plausible hypothesis is that the interactions between carbon or nitrogen and the transition-metal ions broaden the *d*-band and separate it into two subbands with a narrow conduction band between. Most of the Fermi levels of the refractory carbides and nitrides lie in the conduction band. Conclusion: This premise is generally not valid.

PREMISE 3. *Carbon donates about one electron and nitrogen two electrons to the d-band of the interstitial compound.*

This premise is part of the transfer theory of Dempsey. The hypothesis for the carbides appears to agree well with most of the properties, including those for nonstoichiometric phases. These properties include susceptibilities,

Hall coefficients, γ values, soft X-ray emission, and others. This hypothesis, however, almost completely fails to account for the properties of the nitrides. While the data are still scarce, a better hypothesis is that nitrogen is negatively charged or that electrons are transferred from the d bands to the sp bands of the nitrogen. Thus, in conclusion, this hypothesis still appears to be valid for the carbides but definitely not for the nitrides.

X. General Observations

The previous discussion of the electrical and magnetic properties of carbides and nitrides suggests the following observations:

(a) The electrical and magnetic properties of carbides and nitrides are similar to those of the parent transition metal elements. The carbides and nitrides are metallic; their resistivities are comparable to those of the transition metals. They are weakly diamagnetic or paramagnetic. The properties of the carbides and nitrides are, however, different from those of the transition metals; and it is not valid to assume that their properties can be explained by simply adjusting the height of the Fermi level in the band structure approximately applicable for the pure transition elements.

(b) The properties of the carbides and nitrides are sensitive to the non-metal-to-metal ratio. These dependencies, while probably very significant, have not been adequately investigated.

(c) As a first approximation stoichiometric carbides can be treated as isoelectronic phases. Likewise, the nitrides may be treated as isoelectronic phases. If one is dealing with properties which depend only upon the density of states at the Fermi level, then one may make the additional assumption that both stoichiometric carbides and nitrides can be treated as isoelectronic phases. The variation of properties is approximately accounted for by adjusting the height of the Fermi level according to the electron concentration. One should remember, however, that the band structures for the carbides and nitrides differ considerably and therefore they are not really isoelectronic.

(d) Even for properties dependent only upon the density of states at the Fermi level, the nonstoichiometric carbides and nitrides cannot be treated as isoelectronic phases despite indications by some physical properties such as magnetic susceptibilities.

(e) Both the VEC and the electron transfer parameter are useful in correlating the electrical, magnetic, and superconducting properties for nearly stoichiometric phases. Only the electron transfer parameter is useful for the nonstoichiometric carbides.

(f) For the carbides, a new band is forming at an electron concentration corresponding to the stoichiometric fourth-group carbides (VEC = 8). The density of states in this band increases monotonically with increasing electron concentration for the stoichiometric phases. This new band is a broad conduction band with a low density of states which separates the bonding and antibonding parts of the band structure. At compositions corresponding to electron concentrations less than VEC = 8 (nonstoichiometric fourth-group carbides) the density of states increases sharply with decreasing electron concentration.

(g) Electrons are transferred from the metal atoms to the $2p$ bands of nitrogen in the nitrides, but in the carbides the transfer is probably in the opposite direction. The degree of transfer may vary with the group number of the transition metal.

(h) Changes in crystal structure which preserve the octrahedral co-ordination polyhedron seem to have little effect on properties which depend upon the density of states at the Fermi level.

References

1. G. V. Samsonov, "High-Temperature Materials, Properties Index." Plenum Press, New York, 1964.
2. E. Rudy and F. Benesovsky, *Planseeber. Pulvermet.* **8**, 72 (1960).
3. P. Costa and R. R. Conte, *in* "Compounds of Interest in Nuclear Reactor Technology" (J. T. Waber, P. Chiotti, and W. N. Miner, eds.), Inst. Metals Div., Spec. Rep. No. 13, p. 3. Edwards, Ann Arbor, Michigan, 1964.
4. E. Friederich and L. Sittig, *Z. Anorg. Allg. Chem.* **144**, 169 (1925).
5. K. Moers, *Z. Anorg. Allg. Chem.* **198**, 262 (1931).
6. J. R. Cooper and R. L. Hansler, *J. Chem. Phys.* **39**, 248 (1963).
7. G. Santoro and R. T. Dolloff, *J. Appl. Phys.* **39**, 2293 (1968).
8. W. S. Williams, *Phys. Rev.* **A135**, 505 (1964).
9. R. G. Lye, *J. Phys. Chem. Solids* **26**, 407 (1964).
10. H. Bittner and H. Goretzki, *Monatsh. Chem.* **93**, 1000 (1962).
11. N. Pessall, J. K. Hulm, and M. S. Walker, Westinghouse Research Laboratories, Final Rep. AF 33 (615)-2729 (1967).
12. J. Piper, *J. Appl. Phys.* **33**, 2394 (1962).
13. B. H. Eckstein and R. Forman, *J. Appl. Phys.* **33**, 82 (1962).
14. W. S. Williams, *Science* **152**, 34 (1966).
15. H. Bell, Y. M. Shy, D. E. Anderson, and L. E. Toth, *J. Appl. Phys.* **39**, 2797 (1968).
16. T. Tsuchida, Y. Nakamura, M. Mekata, J. Sakurai, and H. Takaki, *J. Phys. Soc. Jap.* **16**, 2453 (1961).
17. J. Piper, *in* "Compounds of Interest in Nuclear Reactor Technology" (J. T. Waber, P. Chiotti, and W. N. Miner, eds.), Inst. Metals Div., Spec. Rep. No. 3, p. 3. Edwards, Ann Arbor, Michigan, 1964.
18. F. Itoh, T. Tsuchida, and H. Takaki, *J. Phys. Soc. Jap.* **19**, 136 (1964).

19. L. B. Dubrovskaya and I. I. Matveyenko, *Phys. Metals Metallogr. (USSR)* **19**, No. 2, 42 (1965).
20. H. Bittner and H. Goretzki, *Monatsh. Chem.* **91**, 616 (1960).
21. H. Bittner, H. Goretzki, F. Benesovsky, and H. Nowotny, *Monatsch. Chem.* **94**, 518 (1963).
22. F. Heiniger, E. Bucher, and J. Muller, *Phys. Kondens. Mater.* **5**, 243 (1966).
23. T. H. Geballe, B. T. Matthias, J. P. Remeika, A. M. Clogston, V. B. Compton, J. P. Maita, and H. J. Williams, *Physics (Long Island City, N.Y.)* **2**, 293 (1966).
24. L. E. Toth, J. Zbasnik, Y. Sato, and W. Gardner, *in* "Anistropy in Single-Crystal Refractory Compounds" (F. W. Vahldiek and S. A. Mersol, eds.), Vol. I, p. 249. Plenum Press, New York, 1968.
25. E. Dempsey, *Phil. Mag.* [8] **8**, 285 (1963).
26. P. Costa, *in* "Anistropy in Single-Crystal Refractory Compounds" (F. W. Vahldiek and S. A. Mersol, eds.), Vol. I, p. 151. Penum Press, New York, 1968.
27. L. E. Toth and J. Zbasnik, *Acta Met.* **16**, 1177 (1968).
28. L. E. Toth, M. Ishikawa, and Y. A. Chang, *Acta Met.* **16**, 1183 (1968).
29. R. G. Lye, J. Phys. Chem. Solids, **26**, 407 (1965).
30. L. G. Radosevich and W. S. Williams, *Phys. Rev.* **181**, 1110 (1969).
31. L. G. Radosevich and W. S. Williams, *Phys. Rev.* **189**, 770 (1969).
32. R. E. Taylor, *J. Amer. Ceram. Soc.* **44**, 525 (1961).
33. R. E. Taylor, *J. Amer. Ceram. Soc.* **45**, 353 (1962).
34. K. Mendelssohn and H. M. Rosenberg, *Solid State Phys.* **12**, 223 (1961).
35. J. Callaway, *Phys. Rev.* **113**, 1046 (1959).
36. W. S. Williams, *J. Am. Ceram. Soc.* **49**, 156 (1966).
37. L. G. Radosevich and W. S. Williams, *J. Am. Ceram. Soc.* **53**, 30 (1970).
38. D. W. Fischer and W. L. Baun, *J. Appl. Phys.* **39**, 4757 (1968).
39. D. W. Fischer and W. L. Baun, *in* "Advances in X-ray Analysis" (G. R. Mallet, M. Fay, and W. M. Mueller, eds.), Vol. 9, p. 329. Plenum Press, New York, 1966.
40. R. G. Lye and E. M. Logothetis, *Phys. Rev.* **147**, 622 (1966).
41. J. E. Holliday, *J. Appl. Phys.* **38**, 4720 (1967).
42. J. E. Holliday, *in* "Advances in X-ray Analysis" (G. R. Mallet, M. Fay, and W. M. Mueller, eds.), Vol. 9, p. 365. Plenum Press, New York, 1966.
43. Y. S. Tyan, L. E. Toth, and Y. A. Chang, *J. Phys. Chem. Solids* **30**, 785 (1969).
44. S. A. Nemnonov, *Fiz. Metal. Metalloved.* **24**, No. 2, 268 (1967).
45. S. A. Nemnonov and K. M. Kolobova, *Fiz. Metal. Metalloved.* **22**, No. 5, 680 (1966).
46. S. A. Nemnonov and L. D. Finkel'shteyn, *Fiz. Metal. Metalloved.* **22**, No. 4, 538 (1966).
47. S. A. Nemnonov, A. Z. Menshikov, K. M. Kolobova, E. Z. Kurymayev, and V. A. Trapenznikov, *Trans. AIME* **245**, 1191 (1969).
48. V. Ern and A. C. Switendick, *Phys. Rev.* **A137**, 1927 (1965).
49. G. V. Samsonov and Ya. S. Umanskiy, "Tverdyye Soyedineniya Tugoplavkikh Metallov." State Sci.-Tech. Lit. Publ. House, Moscow, 1957; for English translation, see *NASA Tech. Transl.* **F-102** (1962).
50. Ya. S. Umanskiy, *Ann. Sect. Anal. Phys.-Chim., Inst. Chim. Gen.* **16**, No. 1, 128 (1943).
51. R. Kiessling, *Met. Rev.* **2**, 77 (1957).
52. H. Bilz, *Z. Phys.* **153**, 338 (1958).

7
Superconducting Properties

I. Introduction

A separate chapter has been devoted to carbides and nitrides as superconductors because of their somewhat unusual properties and because of extensive research on the subject. For many applications of superconductors, such as generation of strong magnetic fields, power transmission, and flux pumps, it is necessary to use materials with high values of superconducting critical temperature T_c, critical magnetic field H_c and critical current density J_c. Many carbides and nitrides have comparatively very large T_c's, H_c's, and J_c's. Table I indicates the relative importance of carbides and nitrides as superconductors and lists some representative phases with high values for T_c and H_c. Most high T_c and H_c phases belong either to the $A15$ group (β-W compounds such as Nb_3Sn and V_3Si) or to the carbide and nitride family. The superconducting properties of the $A15$ compounds are, in general, superior to those of the carbides and nitrides, but the difference is not exceedingly large.

II. Superconducting Critical Temperatures

Table II lists representative T_c values for binary carbides and nitrides at the composition closest to the stoichiometric, although in some cases, such

215

as VC, there is an appreciable deviation from the stoichiometric composition. When available, T_c's derived from heat capacity measurements were used. The values chosen are probably the most reliable at the present time; however, a number of factors make this assessment difficult, including metal-

TABLE I

REPRESENTATIVE SUPERCONDUCTING MATERIALS WITH HIGH T_c's AND H_c's

Phase	Structure	T_c (°K)	H_{c_2} (kOe)
$Nb_3(Al_{0.8}Ge_{0.2})$	$A15$	20.05 (a)	—
Nb_3Sn	$A15$	18.05 (b)	221 at 4.2°K (c)
Nb_3Al	$A15$	18.0 (b)	—
(NbN–TiC)	$B1$	18.0 max (d)	120 max at 4.2°K (d)
$NbN_{0.7}C_{0.3}$	$B1$	17.8 max (e)	132 max at 4.2°K (e)
V_3Si	$A15$	17.1 (b)	235 extrapolated to 0°K (f)
(NbN–TiN)	$B1$	17.0 (d, g)	>140 at 4.2°K (g)
V_3Ga	$A15$	16.5 (b)	210 extrapolated to 0°K (f)
NbN	$B1$	17.3 (e)	132 at 4.2°K (h)
Nb_3Ga	$A15$	14.5 (b)	—
MoC	$B1$	14.3 (i)	98.5 at 1.2°K (j)
MoN	—	12.0 (b)	—
Mo_3Al_2C	β-Mn	10.0 (k)	156 at 1.2°K (j)
Nb–50% Zr	bcc	9.5 (l)	115 at 1.2°K (m)
Nb–60% Ti	bcc	8.7 (l)	145 at 1.2°K (m)

References for Table I

a. B. T. Matthias, T. H. Geballe, L. D. Longinotti, E. Corenzwit, G. W. Hull, R. H. Willens, and J. P. Maita, *Science* **156**, 645 (1967).
b. Compiled by B. T. Matthias, T. H. Geballe, and V. B. Compton, *Rev. Mod. Phys.* **35**, 1 (1963).
c. D. B. Montgomery and W. Sampson, *Appl. Phys. Lett.* **6**, 108 (1965).
d. N. Pessall, J. K. Hulm, and M. S. Walker, Westinghouse Research Laboratories, Final Rep. AF 33 (615)-2729 (1967).
e. M. W. Williams, K. M. Ralls, and M. R. Pickus, *J. Phys. Chem. Solids* **28**, 333 (1967).
f. E. Saur and H. Wizgall, "Les champs magnetiques intenses." Colloq. Int., Grenoble, 1966.
g. C. M. Yen, L. E. Toth, Y. M. Shy, D. E. Anderson, and L. G. Rosner, *J. Appl. Phys.* **38**, 2268 (1967).
h. K. Hechler, E. Saur, and H. Wizgall, *Z. Phys.* **205**, 400 (1967).
i. R. H. Willens, E. Buehler, and B. T. Matthias, *Phys. Rev.* **159**, 327 (1967).
j. H. J. Fink, A. C. Thorsen, E. Parker, V. F. Zackay, and L. E. Toth, *Phys. Rev.* **138 A** 1170 (1965).
k. J. Johnston, L. E. Toth, K. Kennedy, and E. R. Parker, *Solid State Commun.* **2**, 123 (1964).
l. J. K. Hulm and R. D. Blaugher, *Phys. Rev.* **123**, 1569 (1961).
m. T. G. Berlincourt and R. R. Hake, *Phys. Rev.* **131**, 140 (1963).

to-nonmetal ratio, impurities, overall composition, crystal structure, and electron concentration. While these are not the only factors affecting T_c's in carbides and nitrides, they are among the most important.

IIA. Effects of Metal-to-Nonmetal Ratio

For many years, literature on the superconducting T_c's of carbides and nitrides was quite contradictory. For example, Meissner and Franz (1) reported that TaC has a T_c of 9.4°K, but other workers reported that tantalum carbides remained normal to as low as 1.2°K. Literature values for niobium carbide were equally confusing, with a range from 6 to 10.5°K. The reported transition range for niobium nitride was from 6 to 24°K. In 1952, Rögener (2) made an important contribution toward clarifying some of the wide variations in reported values for the superconducting critical temperatures for these phases by showing that the T_c's of NbN_{1-x} are highly sensitive to the nitrogen-to-niobium ratio. Subsequent investigations have shown that the T_c's of NbC_{1-x} (3, 4), TaC_{1-x} (3, 4), VN_{1-x} (5), TiN_{1-x} (5), and HfN_{1-x} (6, 7), systems are also highly sensitive to the metal-to-nonmetal ratio.

Figures 1–3 systematize the T_c variations with composition for the B1-structured monocarbides and nitrides (3–10). For most systems T_c decreases sharply with decreasing carbon or nitrogen content. The highest T_c is associated with the stoichiometric composition.

The compound HfN_{1-x} has an unusual T_c variation. At nitrogen-deficient compositions, T_c is relatively insensitive to the N/Hf ratio, but at nitrogen-rich compositions, T_c rapidly decreases as the N/Hf ratio exceeds one. The HfN system is the only system for which the T_c variation has been studied at nonmetal-to-metal ratios exceeding one.

The T_c variation in the NbN_{1-x} system is also unusual, since T_c may not peak at the stoichiometric composition (8). There appears, however, to be some discrepancy in the data on this system (9, 10). Difficulties in obtaining accurate chemical analyses could account for these differences.

The strong dependence of T_c on the nonmetal-to-metal ratio cannot be explained by the usual correlations between favorable electrons-per-atom ratio and T_c's, as originally proposed by Matthias (3, 11). Giorgi and co-workers (6), however, did discover an interesting correlation between lattice parameter and T_c. They found that when T_c is plotted as a function of lattice parameter, T_c always increases as lattice parameter increases for all B1-structured monocarbides and mononitrides.

It is likely that defect structure plays a significant role in controlling T_c. Doyle et al. (12) have shown that the T_c of stoichiometric TiO with the B1

TABLE II

SUPERCONDUCTING CRITICAL TEMPERATURES OF BINARY CARBIDES IN
GROUPS IV–VI AND REPORTED STRUCTURES [a,b]

IV	V	VI
TiC—B1—<0.05°K (a)	V_2C—L_3'—<1.20°K (c)	$Cr_{23}C_6$—complex fcc—<1.20°K (c)
	$VC_{0.88}$—B1—<0.05°K (a)	Cr_3C_2—ortho—<1.20°K (c)
		Cr_7C_3—trigonal—1.20°K (c)
ZrC—B1—<0.05°K (a)	$NbC_{0.48}$—ε-Fe_2N—<1.5°K (d)	Mo_2C—orthorhombic—4.05°K (f)
	NbC—B1—11.1°K (e)	Mo_2C—L_3'—2.8°K (f)
		Mo_3C_2—hexagonal—8.0°K (f)
		α-MoC—B1—14.3°K (e)
HfC—B1—<1.68°K (b)	$TaC_{0.47}$—C6—<1.75°K (d)	W_2C—L_3'—2.74°K (g)
	TaC—B1—10.35°K (a)	WC—B1—10.0°K (e)
		WC—B_h—<1.28°K (g)

TABLE IIA

SUPERCONDUCTING CRITICAL TEMPERATURES OF BINARY NITRIDES IN GROUPS III–IV AND REPORTED STRUCTURES [b]

III	IV	V	VI
ScN—B1—<1.4°K	(h) TiN—B1—5.49°K (a)	VN—B1—8.5°K (a)	CrN—B1—<1.28°K (g)
YN—B1—<1.4°K (i)	ZrN—B1—10.0°K (a)	V$_2$N—hex—<1.20°K (c)	
		Nb$_2$N—hex—<1.20°K (c)	Mo$_2$N—fcc—5.0°K (g)
		NbN$_{0.8}$—tetra—8.90°K (k)	MoN—hex—12.0°K (g)
		NbN$_{1-x}$—B1—17.3°K (l)	
		NbN—B$_x$—<1.94°K (m)	
LaN—B1—<1.8°K (j)	HfN—B1—8.83°K (a)	Ta$_2$N—hex—<1.20°K (c)	W$_2$N—fcc—<1.28°K (g)
		TaN—B35—<1.20°K (c)	

^{a} The exact crystal structure of the Me$_2$C phase is, in many cases, uncertain because at the time of the measurement the structures had not been determined by neutron diffraction. The values reported refer to the crystal structure reported and not necessarily to the structure later determined by neutron diffraction.

^{b} Compositions are only approximate.

References for Table II

a. N. Pessall, J. K. Hulm, and M. S. Walker, Westinghouse Research Laboratories, Final Rep. AF 33 (615)-2729 (1967).

b. Y. S. Tyan and L. E. Toth, unpublished results (1969).

c. G. F. Hardy and J. K. Hulm, *Phys. Rev.* **93**, 1004 (1954).

d. L. E. Toth, M. Ishikawa, and Y. A. Chang, *Acta Met.* **16**, 1183 (1968).

e. R. H. Willens, E. Buehler, and B. T. Matthias, *Phys. Rev.* **159**, 327 (1967).

f. L. E. Toth and J. Zbasnik, *Acta Met.* **16**, 1177 (1968).

g. B. T. Matthias and J. K. Hulm, *Phys. Rev.* **87**, 799 (1952).

h. W. Meissner and H. Franz, *Z. Phys.* **65**, 30 (1930); *Naturwissenschaften* **18**, 418 (1930).

i. L. E. Toth, C. P. Wang, and C. M. Yen, *Acta Met.* **14**, 1403 (1966).

j. W. T. Ziegler and R. A. Young, *Phys. Rev.* **90**, 115 (1953).

k. G. Lautz and E. Schröder, *Z. Naturforsch.* **A11**, 517 (1956).

l. M. W. Williams, K. M. Ralls, and M. R. Pickus, *J. Phys. Chem. Solids* **28**, 333 (1967).

m. E. Schröder, *Z. Naturforsch.* **A12**, 247 (1957).

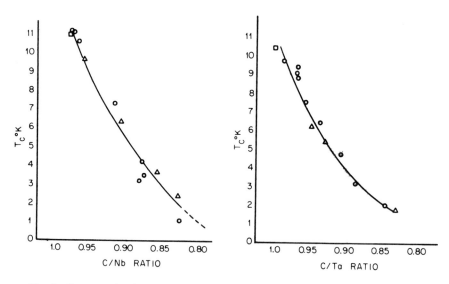

Fig. 1. Superconducting transition temperatures in the NbC_{1-x} and TaC_{1-x} binary systems decrease sharply with decreasing C/Nb and C/Ta ratios. $\circ$: data from Giorgi *et al.* (*3*); $\triangle$: data from Toth *et al.* (*4*); $\square$: data from Pessall *et al.* (*10*).

structure depends strongly on the concentration of lattice vacancies. At the stoichiometric composition of this material, 15% of both sublattices are vacant. By heating TiO to 1650°C and applying pressure up to 90 kbars,

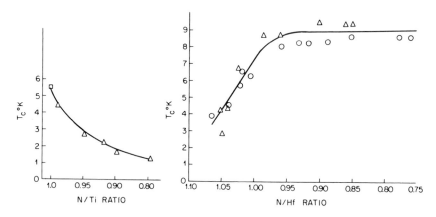

Fig. 2. Variation of the superconducting transition temperature in the TiN_{1-x} and HfN_{1-x} binary systems. Left: $\square$: data from Pessall *et al.* (*10*); $\triangle$: data from Toth *et al.* (*5*). Right: $\circ$: data from Giorgi *et al.* (*6*); $\triangle$: data from Lee (*7*).

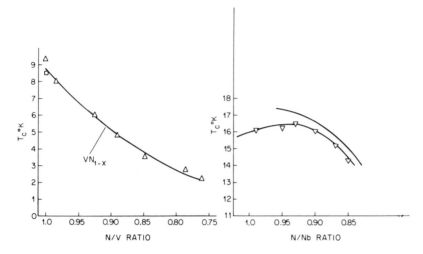

Fig. 3. Variation of the superconducting transition temperature in the VN_{1-x} and NbN_{1-x} binary systems. △ : data from Toth *et al.* (*5*) ; △ : data from Horn and Saur (*8*) ; — : data from Williams *et al.* (*9*) ; □ : data from Pessall *et al.* (*10*).

they were able to reduce the vacancy concentration to nearly zero and to increase T_c from 0.7 to 2.3°K. Figure 4 shows the dependence of T_c on the vacancy concentration of TiO.

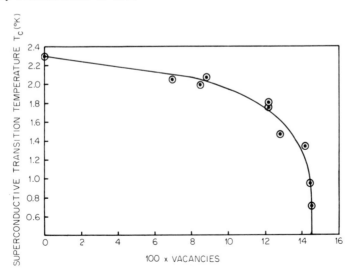

Fig. 4. In titanium monoxide, the superconducting T_c is a sensitive function of the vacancy concentration. [From Doyle *et al.*, *Phys. Lett.* **A26**, 604 (1968).]

No similar studies have been made for carbides and nitrides, and it is not yet possible to separate the effects of composition from the effects of defect structure. It seems likely, however, that the presence of vacancies on either sublattice would tend to reduce T_c. Even at the stoichiometric compositions, several nitrides have appreciable vacancy concentrations that could affect T_c. Since T_c tends to decrease with decreasing nonmetal content at substoichiometric compositions in most systems, and since T_c is probably dependent on vacancy concentration even at the stoichiometric compositions, studies at very high pressures and temperatures that would produce stoichiometric compositions and low vacancy concentrations might result in T_c's higher than those currently obtained for carbides and nitrides.

The effect of the metal-to-nonmetal ratio on T_c has been investigated only for the $B1$ phase. Generally the subcarbides and nitrides have lower T_c's than the monocarbides, and so they have not been studied extensively. Sadogapan and Gatos (13), however, have reported an unusually high T_c for orthorhombic Mo_2C, 12.2°K. They suggest that the value of 2.78°K reported by Matthias and Hulm (14) is low because of a low carbon-to-metal ratio. Subsequent measurements by Toth and Zbasnik (15) on orthorhombic Mo_2C with excess carbon present yielded a T_c of 4.05°K. Morton (16) obtained a T_c greater than 4.2°K by carbonizing molybdenum wire in xylene at temperatures between 1400 and 2100°C. Apparently the T_c of Mo_2C is very sensitive to methods of preparation and carbon content, but no careful analysis of this great variation in T_c is available.

IIB. Effects of Impurities on T_c

A principal impurity affecting the T_c's of carbides and nitrides is oxygen. Because of high reaction temperatures, oxygen is difficult to eliminate completely, especially for the fourth-group transition elements. Oxygen generally lowers the T_c of carbides and nitrides. Hardy and Hulm (17) were the first to report this effect in VN and TiN. Similar results have been reported for NbN_{1-x} (15). The T_c of NbN_{1-x}, for example, is reduced from 15 to about 10°K by the addition of oxygen in solid solution (10).

It is likely that the manner in which T_c is affected by oxygen depends on the metal-to-nonmetal ratio of the nitride or carbide. This ratio affects the activities of the elements and the limits of solid solubility for oxygen. Unfortunately the effect of oxygen on T_c as a function of metal-to-nonmetal ratio has not been systematically investigated. It is possible, however, that some of the differences in the variation of T_c with the metal-to-nonmetal ratio in HfN_{1-x} are attributable in part to a varying oxygen content. The T_c of these samples is highly sensitive to the manner of sample preparation

(7). Nitrogen deficient samples heated in a vacuum (1×10^{-5} Torr) to 1650°C have substantially lower T_c's than samples with the same N/Hf ratio which were prepared by nitriding at different partial pressures and temperatures or by reacting Hf/HfN mixtures in purified He or Ar at 1350°C (Fig. 5).

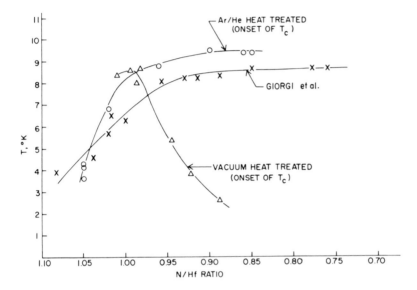

Fig. 5. Superconducting T_c's of substoichiometric HfN_{1-x} heated in vacuum are lower than those for the same N/Hf ratio heated in an equilibrium N_2 atmosphere [Giorgi et al. (6)] or an Ar–He gas. The T_c lowering may be due to oxygen contamination. The T_c's of the Ar–He treated samples were broad indicating that the samples were not homogeneous. [After Lee (7).]

The nitrogen-deficient samples tend to oxidize in the relatively poor vacuum, causing the T_c to be reduced. The nearly stoichiometric samples are not affected as much by the vacuum treatment, presumably because of the reduced activity of the hafnium. The T_c's of samples with initial compositions of N/Hf > 1 increase after vacuum heat treatment at 1650°C because of loss of nitrogen and change of composition to one more nearly stoichiometric. While the differences in T_c may be attributable to a varying oxygen content, other factors, such as the defect concentration, are also affected by the heat treatment and these factors, in turn, affect T_c. Unfortunately, reliable oxygen analyses of these samples are difficult to obtain and therefore it is not possible to separate the effects of oxygen contamination from those of vacancy defect concentration in this system.

The lattice parameters of these samples also show a surprising dependence on heat treatment; samples closest to stoichiometry are the most affected (Fig. 6). No explanation has been found for this behavior, but it seems likely that defect structure is also sensitive to the method of heat treatments.

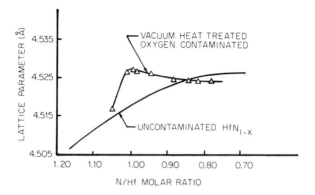

Fig. 6. Lattice parameters of HfN_{1-x} are sensitive to the method of heat treatment and impurity level. Vacuum heat treating (1×10^{-5} Torr at 1650°C) increases the oxygen content. Although the effect of this oxygen on lattice parameter is most pronounced for nearly stoichiometric compositions, the effect on T_c is most pronounced at low N/Hf ratios. [Data after Lee (7); consult Chapter 3, Fig. 13 for further details on the lattice parameter.]

IIC. T_c's in Ternary and Higher-Order Systems

Because monocarbides and nitrides have very high T_c's, several investigators have studied ternary and higher order systems in an attempt to increase T_c. Generally, these studies have been performed on nearly stoichiometric samples because of the tendency for T_c to decrease with decreasing nonmetal-to-metal ratio. The highest T_c's usually found in these systems have been about 18°K. These occur in pseudobinaries of NbN with NbC (10, 18, 19) and TiC (10). Most of these high T_c compounds have nonmetal-to-metal ratios substantially less than one. Therefore T_c's > 18°K could reasonably be expected of stoichiometry were more closely approached. Studies by Williams et al. (9) on nitriding NbC_xN_y under various pressures greater than one atmosphere tend to confirm this expectation: the higher the nitriding pressure, the higher the T_c. According to the thermodynamic discussion of the Wagner–Schottky model in Chapter 4, it appears that optimum conditions for increasing T_c beyond 18°K would be low nitriding temperatures and high nitriding pressures.

Figures 7–10 summarize the T_c variations with composition in several ternary carbide–nitride systems. Additional information may be found by consulting references (*8–10, 18–24*).

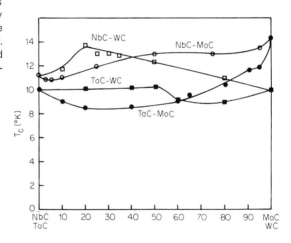

Fig. 7. Superconducting T_c's of solid solutions of group V and VI monocarbides with the (*B*1) crystal structure show no pronounced maxima as a function of composition. Some of the small maxima, as in the NbC–WC system, may be due to variations in the combined carbon content. These alloys were prepared by splat-cooling. [After Willens *et al.* (*24*).]

Figure 7 shows that the maximum T_c in carbide pseudobinaries is 14.3°K and occurs for pure MoC_{1-x}. There is only one pronounced maximum in the composition dependence of T_c, that for the NbC–WC binary system.

For the nitride pseudobinaries a pronounced maximum exists in the NbN–TiN system (Fig. 8). Chemical analysis of the nitrogen content reported by Yen *et al.* (*22*) shows that part of the reason for the maximum is decreasing nitrogen content with increasing NbN content.

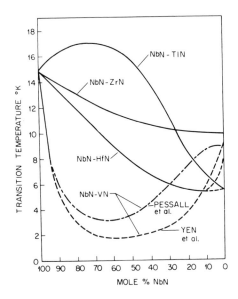

Fig. 8. Approximate variation of T_c with composition for several pseudobinary nitride systems. [After (*8, 10, 20–23*).]

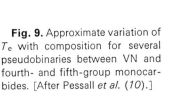

Fig. 9. Approximate variation of T_c with composition for several pseudobinaries between VN and fourth- and fifth-group monocarbides. [After Pessall *et al.* (*10*).]

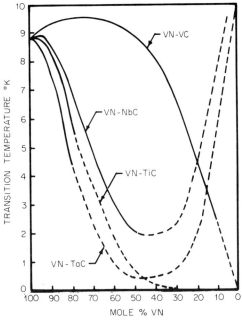

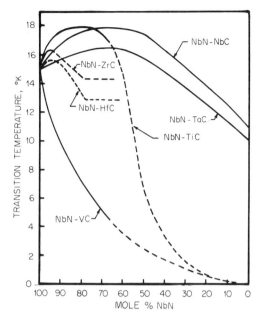

Fig. 10. Approximate variation of T_c with composition for several pseudobinary systems between NbN and fourth- and fifth-group monocarbides. [After Pessall *et al. J. Phys. Chem. Solids* **29**, 19 (1968). Reprinted by permission of Pergamon Press.]

In the NbN–VN (Fig. 8), the pronounced minimum in T_c cannot be explained by a variable nitrogen content. Similar minima are found in several pseudobinaries involving VN and VC with the other monocarbides and nitrides (Fig. 9). No explanation has been offered for these minima.

Several important high T_c maxima exist in the NbN–NbC and NbN–TiC pseudobinaries (Fig. 10). The unknown variation of the nonmetal-to-metal ratio with the mole fraction of NbN may partly explain these maxima.

IID. Effect of Crystal Structure on T_c

Most investigations on superconductivity of carbides and nitrides have centered on the high T_c B1-type compounds. The only other crystal structure associated with very high T_c's is the cubic A15 type of the β-W compounds. A number of investigators have noted unusually high T_c's for carbides and nitrides which do not have the B1 structure (*13, 25–29*), and they have suggested that the octahedral coordination polyhedron common to all these crystal structures is responsible for the high T_c's (*13, 25–29*). (For a discussion of the octahedral coordination polyhedron and the crystal structure types, see Chapter 2.)

The influence of crystal structure on T_c is best illustrated in the MoC family of compounds (Table III) (*29*). Their T_c's are generally the highest

TABLE III

SUPERCONDUCTING CRITICAL TEMPERATURES OF MOLYBDENUM CARBIDES [a]

Phase	Crystal structure	Transition temperature (°K)	Ref.
$MoC_{1.0}$	$B1$ (cubic)	14.3 [b]	(a)
$MoC_{0.67}$	$B1$ (cubic)	12.2 [c, d]	(b)
Mo_3Al_2C	β-Mn (cubic)	10.0 [d]	(c)
Mo_3C_2	Complex (hexagonal)	9.0 [c, d], 8.0 [d]	(b, d)
Mo_2GaC	H-phase (hexagonal)	4.1–3.7	(e)
Mo_2BC	Orthorhombic	7.0–5.3	(f)
Mo_2C	$L_3{}'$	2.78	(g)
Mo_2C	Orthorhombic	4.05 [e], 12.2	(d, h)
$Mo_{4.8}Si_3C_{0.6}$	$D8_8$ (hexagonal)	7.6	(h)

[a] After Toth (29). [Courtesy of the *Journal of Less-Common Metals*.]
[b] Refers to beginning of a broad transition of a splat cooled material.
[c] The T_c is sensitive to the severity of quench; see Toth *et al.* (25) for details.
[d] Refers to midpoint of transition.
[e] Refers to low temperature heat capacity data.

References for Table III

a. R. H. Willens and E. Beuhler, *Appl. Phys. Lett.* **7**, 25 (1965).
b. L. E. Toth, E. Rudy, J. Johnston, and E. R. Parker, *J. Phys. Chem. Solids* **26**, 517 (1965).
c. J. Johnston, L. Toth, K. Kennedy, and E. R. Parker, *Solid State Commun.* **2**, 123 (1964).
d. L. E. Toth and J. Zbasnik, *Acta Met.* **16**, 1177 (1968).
e. L. E. Toth, W. Jeitschko, and C. M. Yen, *J. Less-Common Metals* **10**, 29 (1966).
f. L. E. Toth. *J. Less-Common Metals* **13**, 129 (1967).
g. B. T. Matthias and J. K. Hulm, *Phys. Rev.* **87**, 799 (1952).
h. V. Sadagopan and H. C. Gatos, *J. Phys. Chem. Solids* **27**, 235 (1966).

yet reported for compounds with similar crystal structures and in some cases for entire symmetry classifications. MoC_{1-x} has the highest T_c for a binary $B1$ structured carbide (30), although not the highest T_c for a compound with this structure. Mo_3Al_2C has the highest T_c for a β-Mn structured compound. The second highest T_c for a compound with a hexagonal structure is found for Mo_3C_2. The highest T_c (12.2°K) (13) for a compound with an orthorhombic structure is found for Mo_2C, and the third highest for Mo_2BC. The compound Mo_2GaC is the only H-phase reported to superconduct. $Mo_{4.8}Si_3C_{0.6}$ has the highest reported T_c for a compound with the $D8_8$ structure and one of the highest T_c's for a hexagonal structure. The above comparisons are based on the compilation of T_c data by Matthias *et al.* (*31*).

Because all members of the molybdenum carbide family have high T_c's regardless of symmetry, composition, and other factors such as electron concentration, the common Mo_6C octahedra are most likely favorable for high T_c's. Other molybdenum carbides with this coordination polyhedron can probably be expected to superconduct.

Sadagopan and Gatos (*13*) were among the first to point out that co-ordination polyhedra are important factors affecting T_c in carbides and nitrides. From knowledge of the superconducting behavior of the $B1$ phase, they believe that a number of complex phases can be expected to super-conduct. Superconductivity is favored in the complex phases when it occurs in the stoichiometric $B1$ phase. They suggest, for example, that stoichiometric Nb_5Si_3C should most likely be a high T_c superconductor because stoichio-metric NbC has a high T_c.

Limited experimental results on complex carbides and nitrides do not in general support the hypothesis of Sadagopan and Gatos (*13*). Table IV lists the T_c's of several carbides and nitrides with complex crystal structures which contain the octahedral coordination polyhedra (*28*). This data, plus the data in Tables II and III, summarize all the known T_c data for carbides and nitrides with crystal structures other than $B1$. According to the above

TABLE IV

H-PHASES AND PHASES WITH FILLED AND ORDERED
β-Mn TYPE TESTED FOR SUPERCONDUCTIVITY [a]

Compound	Type	Transition temperature ($^\circ$K)
Mo_3Al_2C	β-Mn	10.0
Nb_3Al_2N	β-Mn	1.3
Ti_2CdC	H	
Zr_2InC	H	
Hf_2InC	H	
Ti_2PbC	H	
Ti_2InN	H	No superconductivity
Hf_2InN	H	above 1.1 $^\circ$K
V_2AlC	H	
Nb_2InC	H	
Nb_2SnC	H	
Cr_2AlC	H	
Cr_2GaC	H	
Mo_2GaC	H	3.9

[a] After Toth *et al.* (*28*). [Courtesy of the *Journal of Less-Common Metals*.]

hypothesis, most of these phases should be high T_c superconductors, but only the molybdenum carbides have high T_c's. Unfortunately, nothing is known about the stoichiometry of these complex phases. Possibly, molybdenum carbides are unique among carbides and nitrides in their superconducting behavior. This behavior may be related to the small effect deviations from stoichiometry have on the T_c of MoC_{1-x} with the $B1$ structure. The $B1$ MoC phase has a very high T_c with a large deviation from stoichiometry.

IIE. Superconducting T_c's of Thin Nitride-Films

Superconducting thin films (1000–8000 Å) of NbN and Nb–Ti–N and Nb–Zr–N ternary alloys have been prepared by reactive sputtering (*32–38*). (See Chapter 1 for a description of this technique.) These films are particularly interesting for a number of thin-film superconducting devices. The superconducting properties of these films are highly sensitive to the manner of film preparation; unfortunately, many of the factors that affect T_c and the other superconducting properties are poorly understood. Gerstenberg and Hall (*32*), for example, used an oil-diffusion pumped and LN baffled cathode sputtering system to prepare NbN with a T_c in the range between 6 and 9°K. By outgassing a similar system to 10^{-11} Torr, Deis *et al.* (*38*) were able to increase T_c to above 15°K. This T_c is nearly that of the bulk samples. Mitzuoka *et al.* (*36*) used a presputtering vacuum of only 10^{-6} Torr, but they also obtained thin films of NbN with a T_c greater than 15°K by using ac asymmetric sputtering. They believe that the reason for the increased T_c over that obtained for films prepared by dc techniques in a similar vacuum is an increased nitrogen content in the films and not necessarily greater purity. In addition to the type of sputtering and the quality of the vacuum of the presputtering system, the superconducting properties are also dependent upon the substrate material, the rate of deposition, the substrate temperature, and the nitrogen partial pressure during sputtering.

Bell *et al.* (*35*) prepared ternary films in the Nb–Ti–N and Nb–Zr–N systems by developing a modified triode sputtering apparatus in which it was possible to vary the Nb/Ti (Nb/Zr) ratio of the films by simply varying the potential on the Nb and Ti(Zr) sputtering targets. The associated vacuum system was processed as an ultrahigh vacuum system (10^{-9} Torr) prior to sputtering (see Chapter 1 for a description of this apparatus). For the Nb–Zr–N system the T_c's of the ternary films were nearly equal to those of the bulk samples, but for the Nb–Ti–N system the T_c's of the ternary alloys were several degrees lower than those of the bulk (see Fig. 11). The latter system is particularly interesting because it has the highest known T_c in bulk for a pure nitride alloy. While the T_c's of these films are lower than

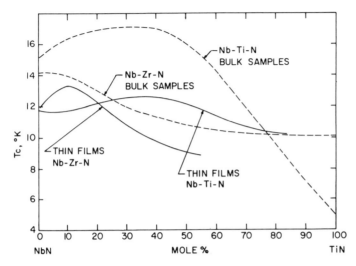

Fig. 11. Superconducting T_c's of thin films of Nb–Ti–N and Nb–Zr–N alloys prepared by Bell *et al.* (*35*) are several degrees lower than the T_c's of bulk samples. The T_c's of the films are highly sensitive to preparation techniques.

those of the bulk, there is no fundamental reason why the T_c's cannot be increased to the bulk values. As discussed in the next section, the upper critical fields and current densities of these films are superior to those of bulk samples. If the T_c's of films of this system could also be increased to at least the T_c's of the bulk, then these films would be very promising for a number of applications in superconducting devices.

III. Critical Magnetic Fields and Critical Currents

Upper critical magnetic fields H_{c2} have been studied extensively on a large number of binary, ternary, and quatenary carbides and nitrides. Several carbides and nitrides have very high values of H_{c2} and are therefore potentially useful for superconducting solenoid applications. Their poor mechanical properties have hindered this development, since normally the solenoids are wound with wires or tapes. Recently, however, the H_{c2} properties of thin nitride films have been found to be superior to those of bulk samples (*33–35, 37, 38*). These films, easily deposited, are ideal for preparing thin film superconducting solenoids economically.

A number of factors are known to affect H_{c2} and current density. These include the nonmetal-to-metal ratio, overall composition in ternary and quatenary systems, impurity content, sample form (bulk or thin film) and orientation of the sample in the magnetic field.

IIIA. Binary Systems—Deviations from Stoichiometry

Table V lists the H_{c_2}'s of several binary carbides and nitrides with the $B1$ structure and of several carbides with complex crystal structures. It is some-

TABLE V

UPPER CRITICAL FIELDS FOR NEARLY STOICHIOMETRIC
CARBIDES AND NITRIDES [a]

Compound	Structure	H_{c_2} (kOe)	Meas. temp. ($^\circ$K)	Ref.
NbN [b]	$B1$	132	4.2	(a–e)
MoC [b]	$B1$	98	4.2	(f, g)
NbC [b]	$B1$	16.9	4.2	(f)
TaC [b]	$B1$	4.6	1.2	(f, d)
ZrN	$B1$	4	4.2	(b, d)
HfN	$B1$	<10 [c]	4.2	(c)
TiN	$B1$	<10 [c]	4.2	(c, d)
VN	$B1$	—	—	—
η-MoC$_{2/3}$	Hex	47	4.2	(g, h)
Mo$_2$C	Hex	30	4.2	(h)
Mo$_2$BC	Ortho	28	4.2	(g)
Mo$_3$Al$_2$C	β-Mn	101	4.2	(g)

[a] Bulk samples.
[b] These phases are not stoichiometric and the stoichiometry affects H_{c_2} (see Fig. 12). Estimates of the stoichiometry are as follows: NbN$_{\sim 1.0}$, MoC$_{0.67}$, NbC$_{0.97}$, TaC$_{0.97}$.
[c] Estimated from variation of H_{c_2} in ternary alloys.

References for Table V

a. T. H. Courtney, J. Reintjes, and J. Wulff, *J. Appl. Phys.* **36**, 660 (1965).
b. L. E. Toth, C. M. Yen, L. G. Rosner, and D. E. Anderson, *J. Phys. Chem. Solids* **27**, 1815 (1966).
c. C. M. Yen, L. E. Toth, Y. M. Shy, D. E. Anderson, and L. G. Rosner, *J. Appl. Phys.* **38**, 2268 (1967).
d. N. Pessall, J. K. Hulm, and M. S. Walker, Westinghouse Research Laboratories, Final Rep. AF 33 (615)-2729 (1967).
e. K. Hechler, E. Saur, and H. Wizgall, Z. *Phys.* **205**, 400 (1967).
f. H. J. Fink, A. C. Thorsen, E. Parker, V. F. Zackay, and L. Toth, *Phys. Rev.* **A138**, 1170 (1965).
g. L. E. Toth and J. Zbasnik, *Acta Met.* **16**, 1177 (1968).
h. N. Morton, *Cryogenics* **8**, 30 (1968).

what surprising that only NbN and MoC have high H_{c2}'s, since the T_c's of NbC, TaC, ZrN, and HfN are not appreciably lower than those of NbN and MoC. The reason for this wide range of H_2 values will be discussed later.

Only one study exists on the effects of stoichiometry deviations on the upper critical field. Williams *et al.* (9) studied the H_{c2} of NbN$_{1-x}$ as a function of nitriding pressure and temperature. Unfortunately, these authors did not report their data in the form H_{c2} versus N/Nb ratio, but rather in the form of H_{c2} versus nitriding pressure, with nitriding temperature and time unspecified. Nevertheless, the information given is sufficient to estimate qualitatively the H_{c2} variation with N/Nb ratio (Fig. 12). The upper critical magnetic field H_{c2} is strongly dependent on the N/Nb ratio and increases as the N/Nb ratio increases.

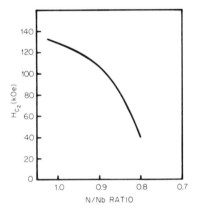

Fig. 12. The upper critical field H_{c_2} increases as the N/Nb ratio increases. [The curve was estimated from the data of Williams *et al.* (9) on the variation of H_{c2} with nitriding pressure and temperature.]

IIIB. Ternary Systems

Figures 13 and 14 show H_{c2} values at 4.2°K for pseudobinaries between NbN and several mononitrides and monocarbides (10, 20–22, 39). The nonmetal-to-metal ratio has been determined for only a few systems, but most samples were prepared so that this ratio was nearly one. There is some doubt about the cause of the maxima in H_{c2} observed at compositions close to NbN. For those systems that were chemically analyzed, the nitrogen content was found to decrease slightly as pure NbN was approached. As Fig. 12 shows, this decreased nitrogen content could result in the observed maxima in H_{c2} in the pseudobinary systems.

The H_{c2} values of a number of pseudoternaries between (NbN)$_{0.75}$, (NbC)$_{0.25-x}$, and (MeC)$_x$ or (MeN)$_x$ have been investigated because of their very high T_c's (*10*). Since H_{c2} values are comparable to those of the pseudo-binary alloys, however, (120 to 130 kOe) using these more complex systems offers no real advantages.

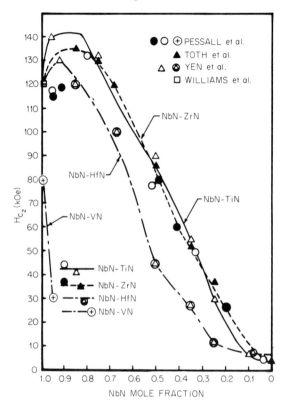

Fig. 13. The variation of H_{c2} with composition in several pseudobinary systems. [After (*10, 20–22*).]

IIIC. Thin Films

Figure 15 shows the H_{c2} values of thin nitride films in the Nb–Ti–N system as a function of composition (*37*). The H_{c2} values were determined in dc magnetic fields with H perpendicular to the film. The films were prepared by reactive sputtering, using the techniques described in Chapter 1. The T_c's of these films are shown in Fig. 11. An unusual feature of the films is that at certain composition ranges their upper critical fields are substantially higher than those shown in Fig. 13 for the sintered Nb–Ti–N samples. The reason for this enhancement in H_{c2} is unclear, since the T_c's of these films are 3–5°K lower than the bulk, and since the H_{c2} values of the relatively thick (2000–8000 Å) films are independent of film thickness. Furthermore,

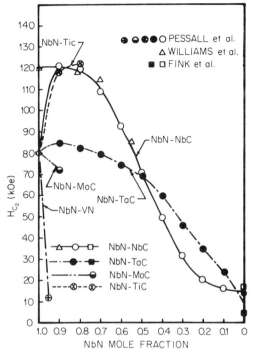

Fig. 14. Variation of the upper critical field with composition in the pseudobinary systems between NbN and several monocarbides. The temperature of measurement was 4.2°K. The nonmetal-to-metal ratio was generally not specified. [After (9, 10, 39).]

the H_{c2} values are relatively independent of film orientation in the field, and there is no possibility of having measured H_{c3}.

Recently Deis et al. (38) measured the H_{c2} properties of NbN films with T_c's greater than 15°K. They found that the H_{c2} values at 4.2°K are greater

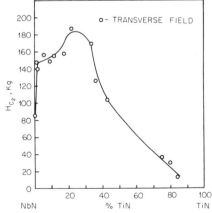

Fig. 15. The upper critical fields of thin films in the Nb–Ti–N ternary system are generally higher than those found for bulk samples of the same composition. The films were measured in dc fields at 4.2°K with H perpendicular to the film. The T_c's of these films are shown in Fig. 11. [After Zbasnik et al. (37).]

than 200 kOe. The measurements were performed in dc magnetic fields with the plane of the film perpendicular to the field.

Several authors (37, 38) have suggested that the reason for the significant enhancement in H_{c2} values of nitride films over those of bulk nitrides is due primarily to the increased normal state resistivity of the films. (See Section IIIE for a discussion of the importance of the normal state resistivity in determining H_{c2}.) The normal state resistivities of the films are several times those of the bulk materials of the same composition. The normal state resistivity is, however, sensitive to the method of preparing the films (36). No detailed studies have been made to correlate resistivity and sputtering conditions with H_{c2} values.

The enhancement of H_{c2} for the nitride films to such a high level is a significant development. Conceivably, with proper sputtering conditions, H_{c2} values for nitride films could appreciably exceed 200 kOe. Deis et al. (38), using an extrapolation technique, estimate H_{c2} values of over 250 kOe for the 15°K NbN films. The difference between these values and those shown in Fig. 15 apparently results from the 2–3°K enhancement in T_c. Increasing the T_c in the Nb–Ti–N alloy films to their bulk values (~ 17°K) should further increase H_{c2} to perhaps over 300 kOe. Increasing T_c values in the films also improves the J_c values. This is one promising area where further experimentation could optimize superconducting properties.

IIID. Critical-Current Densities

Critical-current densities J_c have been measured on sintered compacts (10, 20–22, 39–41), nitrided wires (42, 43), single crystals (44), and thin nitride films (33–35, 37, 38, 45). Critical currents are highly sensitive to the method of preparation; values for samples of supposedly identical compositions prepared by different techniques differ by as much as three orders of magnitude. In general, the worst critical-current densities are observed for sintered bulk samples. These current densities are not corrected for sample porosity; no doubt the low values of J_c reflect this porosity. Figure 16 shows the J_c–H characteristics for the Nb–Zr–N system of sintered samples; these characteristics are typical of most other sintered systems. Much higher critical-current densities are observed in nitrided wires, single crystals, and thin films. Figure 17 shows the J_c–H characteristics of thin films of Nb–Ti–N with T_c's shown in Fig. 11 (37). By increasing T_c to 15°K, the much better J_c–H characteristics shown in Fig. 18 are obtained (38). These characteristics were determined in dc fields with H perpendicular to the film and to J. The usual 10-μV criterion was used to determine J_c. The high J_c values

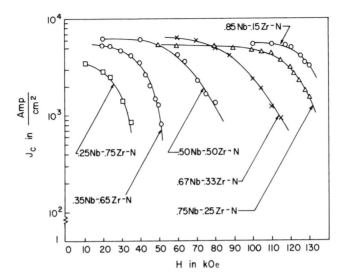

Fig. 16. J_c's for sintered nitride and carbide samples are typically several orders of magnitude lower than those obtainable with thin films, nitrided wires, and single crystals. [After Toth *et al. J. Phys. Chem. Solids* **27**, 1815 (1966). Reprinted by permission of Pergamon Press.]

for the films are comparable at high fields to those of other very high-field superconductors such as Nb_3Sn or V_3Ga.

Critical-current density values depend upon the orientation of the sample in the magnetic field. Some rather unusual "peak" effects have been observed in the J_c–H characteristics of Mo_3Al_2C (*39*) and NbN (Fig. 19) (*43, 44, 46, 47*). Values of J_c for thin films are relatively insensitive to film orientation, although higher J_c values are observed when H is parallel to the film.

III E. Theoretical

A number of investigators (*10, 15, 20, 39, 43, 44, 48*) have analyzed the upper critical magnetic fields of carbides and nitrides in terms of the GLAG[1] theories for hard superconductors [e.g., Maki (*49*) and Werthamer, Helfand, and Hohenberg (WHH) (*50*)]. The GLAG theory predicts that H_{c2} is given by the expression

$$H_{c2} = 319\frac{\rho_0\gamma T_c}{V_m} + 6.85\frac{\gamma^2 T_c^2}{\sqrt{V_m}} \quad \text{in Gauss,} \tag{1}$$

[1] The GLAG (Ginburg–Landau–Abrikosov–Gor'kov) theory is discussed by several authors in *Rev. Mod. Phys.* (January, 1964).

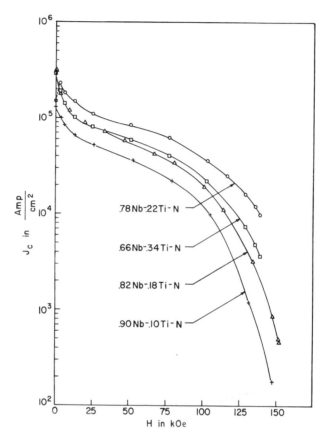

Fig. 17. Superconducting J_c–H characteristics of thin films (2000 Å) of Nb–Ti–N alloys are significantly better than those characteristics obtained with sintered samples of the same composition. The characteristics shown are for H perpendicular to the film and to J. [After Zbasnik *et al.* (*37*).]

where γ is the electronic specific heat in millijoules per mole per degree squared, ρ_0 is the residual resistivity in the normal state, and V_m is the molar volume (*10*). The second term is subject to certain simplifying assumptions about the ratio of the area of the Fermi surface to that for free electrons. In the first term, the coherence length is determined by the electron mean free path, whereas in the second term, the coherence length is determined entirely by the electronic properties of the Fermi surface.

For carbides and nitrides, the first term in Eq. (1) is predominantly responsible for the high H_{c2} values (*10, 15, 20*). Table VI compares observed H_{c2} values with calculated values obtained from experimental determinations

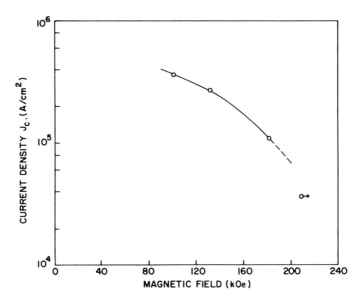

Fig. 18. Excellent superconducting J_c–H characteristics have been observed for NbN thin films at 4.2°K in dc fields. Superconductivity could not be quenched by fields of 200 kOe [After Deis *et al.* (38).]

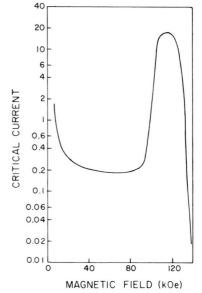

Fig. 19. An unusual peak occurs in the J_c values of nitrided Nb wires at high magnetic fields. A similar effect occurs in the J_c–H characteristics of sintered Mo_3Al_2C samples. [After Hechler *et al.* (*43*).]

TABLE VI

COMPARISON OF OBSERVED AND CALCULATED H_{c_2} VALUES

Phase	Structure	H_{c_2} (kOe) cal.	H_{c_2} (kOe) obs.	Ref.
α-MoC	$B1$	73	98	(a)
η-MoC	Hex	50	47	(a)
Mo_2BC	Ortho	32	28	(a)
Mo_3Al_2C	β-Mn	127	110	(a)
NbC	$B1$	16	16.9	(a, b)
TaC	$B1$	7	4.6	(a, b)
NbN	$B1$	87	80	(c)
$NbN_{0.9}C_{0.1}$	$B1$	88	121	(c)
$NbN_{0.7}C_{0.3}$	$B1$	78	109	(c)
$NbN_{0.5}C_{0.5}$	$B1$	70	71	(c)
$NbN_{0.3}C_{0.7}$	$B1$	46	32	(c)
$NbN_{0.1}C_{0.9}$	$B1$	24	16	(c)

References for Table VI

a. L. E. Toth and J. Zbasnik, *Acta Met.* **16,** 1177 (1968).
b. H. J. Fink, A. C. Thorsen, E. Parker, V. F. Zackay, and L. Toth, *Phys. Rev.* **138,** 1170 (1965).
c. N. Pessall, J. K. Hulm, and M. S. Walker, Westinghouse Research Laboratories, Final Rep. AF 33 (615)-2729 (1967).

of ρ_0, γ, T_c, and V_m. In general, the agreement is fair, although the experiments are difficult because of problems in obtaining accurate ρ_0 values on sintered samples with a high residual porosity. Typically, the calculated ratio of the first and second terms in Eq. (1) falls in the range 10–100, a clear indication that H_{c_2} is limited by the electron mean free path in carbides and nitrides.

For sintered samples of carbides and nitrides, the Maki paramagnetic limitation parameter α is small (*10, 15, 43*). Furthermore, the spin–orbit scattering frequency parameter λ_{so} of the WHH theory is relatively large (*48*). The small α and large λ_{so} implies that paramagnetic corrections to Eq. (1) are not important and that the observed H_{c_2} values should nearly equal those calculated with this equation. For thin films, however, paramagnetic effects may seriously limit H_{c_2} (*38*).

These experiments on factors limiting H_{c_2} are significant because they explain the unusually large variations in H_{c_2} values with composition observed in carbides and nitrides (Figs. 13 and 14), and also the reason for the enhancement of H_{c_2} values for thin nitride films. The H_{c_2} values of fifth-group carbides and fourth-group nitrides are much lower than those of NbN-rich alloys, not because of their lower T_c's, but principally because of their lower γ and ρ_0 values. The thin films have higher H_{c2} values than those of the bulk primarily because of the greatly increased resistivities of the films (37, 38). In seeking high H_{c2} phases among carbides and nitrides, it is, therefore, not only necessary to have large T_c's, but also large γ and ρ_0 values.

IV. Basic Mechanisms

Several investigators have attempted to understand the reasons for the very high T_c's of carbides and nitrides. The T_c is governed by three factors: the "band structure" density of electron states $N(0)$, the electron–phonon interaction parameter V_{ph} and the Debye temperature θ_D. These factors are related to T_c by the strong coupling McMillan (51) formula:

$$\frac{1.45 T_c}{\theta_D} = \exp - \frac{1.04(1 + N(0)V_{ph})}{N(0)\,[V_{ph} - U_c(1 + 0.62 N(0)V_{ph})]} \tag{2}$$

Here U_c is the screened coulomb interaction parameter, and it is usually assumed that $N(0)U_c = 0.1$; this assumption is not critical to the analysis. The larger $N(0)$, V_{ph}, and θ_D, the greater is T_c. $N(0)$ differs from γ and the free electron density of states N_γ in that the enhancement of electron–phonon contributions to the heat capacity are subtracted. Therefore $N(0)$ is a better measure of the real density of electron states than is N_γ. The difference between $N(0)$ and N_γ is given by the expressions

$$N(0) = \frac{N_\gamma}{1 + N(0)V_{ph}} \tag{3}$$

$$N_\gamma = \frac{3\gamma}{2\pi^2 k^2 N_0} \tag{4}$$

where N_0 is Avogadro's number.

Thus by measuring low-temperature specific heats and obtaining γ, θ_D, and T_c, and by applying Eqs. (2)–(4) one determines $N(0)$ and V_{ph}.

Applied to carbides and nitrides, this analysis shows that their very high T_c's result primarily from relatively large values of both V_{ph} and θ_D (4, 10, 15, 20, 52). This behavior is in sharp contrast to the A15 compounds, which derive their high T_c's from large $N(0)$ values. Figure 20 compares the $N(0)$ – V_{ph} values of carbides and nitrides with those for several high-T_c A15 compounds.

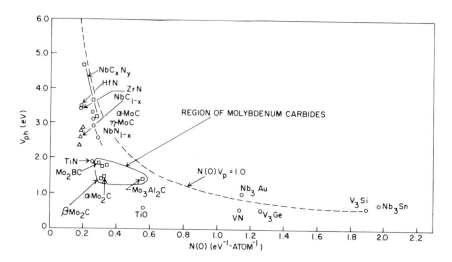

Fig. 20. A plot of V_{ph} versus $N(0)$ suggests that the superconducting properties of molybdenum carbides are closely related to each other. For carbides and nitrides in general, the V_{ph} values are relatively high and the $N(0)$ values relatively low when compared to other materials. With the exception of molybdenum and niobium carbides (*4, 15*). The data were taken from Pessall *et al.* (*10*). [The figure is after Toth and Zbasnik, *Acta Met.* **16,** 1177 (1966). Reprinted by permission of Pergamon Press.]

This type of analysis has also been applied to explain the unusually high T_c's of the molybdenum carbide family of compounds (*15*) (Section IID). As illustrated in Fig. 20, the $N(0) - V_{ph}$ characteristics of the molybdenum carbides are surprisingly similar to each other, despite large changes in crystal structure, crystal symmetry, and even chemical composition. Since all molybdenum carbides are interrelated by a common Mo_6C octahedral coordination polyhedron, we can conclude that structurally related compounds with similar compositions tend to have similar superconducting properties.

This analysis has also been used to evaluate the mechanism responsible for the rapid decrease in T_c with increasing deviations from stoichiometry in the $B1$ phase NbC_{1-x} and TaC_{1-x} (see Fig. 1) (*4*). Figure 21 shows that the parameters $N(0)$, V_{ph}, and θ_D decrease with carbon content in NbC_{1-x} and TaC_{1-x}. Because of the exponential dependence of T_c on V_{ph} and $N(0)$, this rapid decrease in T_c is due primarily to the relatively small decreases in both $N(0)$ and V_{ph}; the decrease in θ_D accounts for only a small part of the decrease in T_c.

There is some indication, as discussed in Chapter 8, that the observed variation in $N(0)$ with composition in NbC_{1-x} and TaC_{1-x} is consistent

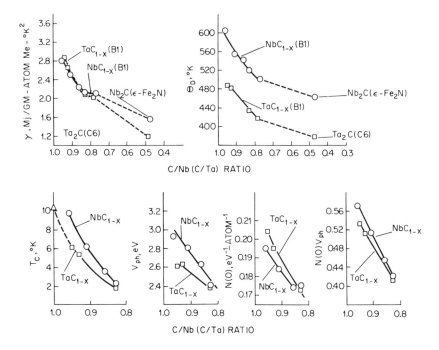

Fig. 21. The rapid decrease in T_c with decreasing C/Nb and C/Ta ratio is accompanied by a decrease in γ and θ_D values. The large change in T_c from about 10°K at stoichiometry to about 2°K at $MeC_{0.8}$ is due primarily to a decrease in $N(0)$ and V_{ph}. [After Toth *et al. Acta Met.* **16**, 1183 (1968). Reprinted by permission of Pergamon Press.]

with theoretical calculations of the electron band structure. The reason for the decrease in V_{ph} with decreasing carbon content, not yet understood, may well be related to the defect structure. Some evidence supports this contention since, as previously pointed out for TiO, the T_c depends on the defect structure.

V. Correlations

Numerous empirical correlations (*6, 10, 23, 53–55*) have been established to relate the superconducting behavior of carbides and nitrides to electronic factors, mass, and volume. The principal use of these correlations has been to systematize the T_c information on carbides and nitrides and to predict, if possible, compositions of new high T_c phases. A number of high T_c phases such as MoC and WC with the $B1$ structure have been found in this manner. None of these correlations can adequately include all the T_c information on binary, ternary, and higher-order systems and also take into account the variations of T_c with stoichiometry deviations.

The optimized distance correlation of Pessall *et al.* (*10, 23*) considers the dilation of the transition metal atoms on forming carbides and nitrides and the electron density. No fundamental significance can be attached to the particular parameters used, although a good correlation is cited.

The valence electron concentration (VEC), first proposed by Bilz (*53*) suggests that T_c depends on the total number of valence electrons (electrons beyond the last completed shell) of both the transition elements and carbon or nitrogen. Carbides and nitrides with the $B1$ crystal structure and with VEC's of 8 have T_c's less than 1°K; those with VEC's of 9 have high T_c's ($\sim 10°$K); and those with VEC's of approximately 10 have very high T_c's ($> 10°$K) (*55*). This correlation suggests that carbides of one group should have superconducting properties similar to those of the nitrides of the previous group. Thus one finds that the fifth-group stoichiometric carbides NbC and TaC have T_c's comparable to fourth-group nitrides ZrN and HfN. The same correlation apparently applies to H_{c2} values and to many other parameters which affect the superconducting behavior. The probable explanation for the success of the correlation is that the VEC also correlates well with the density of electron states as determined by measurements of γ. The VEC correlation will be discussed further in Chapter 8.

References

1. W. Meissner and H. Franz, *Z. Phys.* **65**, 30 (1930).
2. H. Rögener, *Z. Phys.* **132**, 446 (1952).
3. A. L. Giorgi, E. G. Szklarz, E. K. Storms, A. L. Bowman, and B. T. Matthias, *Phys. Rev.* **125**, 837 (1962).
4. L. E. Toth, M. Ishikawa, and Y. A. Chang, *Acta Met.* **16**, 1183 (1968).
5. L. E. Toth, C. P. Wang, and C. M. Yen, *Acta Met.* **14**, 1403 (1966).
6. A. L. Giorgi, E. G. Szklarz, and T. C. Wallace, *Proc. Brit. Ceram. Soc.* **10**, 183 (1968).
7. C. F. Lee, M.S. Thesis, University of Minnesota, 1969.
8. G. Horn and E. Saur, *Z. Phys.* **210**, 70 (1968).
9. M. W. Williams, K. M. Ralls, and M. R. Pickus, *J. Phys. Chem. Solids* **28**, 333 (1967).
10. N. Pessall, R. E. Gold, and H. A. Johansen, *J. Phys. Chem. Solids* **29**, 19 (1968); see also N. Pessall, J. K. Hulm, and M. S. Walker, Westinghouse Research Laboratories, Final Rep. AF 33 (615)-2729 (1967).
11. B. T. Matthias, *Prog. Low Temp. Phys.* **2**, 138 (1957).
12. N. J. Doyle, J. K. Hulm, C. K. Jones, R. C. Miller, and A. Taylor, *Phys. Lett.* **A26**, 604 (1968).

13. V. Sadagopan and H. C. Gatos, *J. Phys. Chem. Solids* **27**, 235 (1966).
14. B. T. Matthias and J. K. Hulm, *Phys. Rev.* **87**, 799 (1952).
15. L. E. Toth and J. Zbasnik, *Acta Met.* **16**, 1177 (1968).
16. N. Morton, *Cryogenics* **8**, 30 (1968).
17. G. F. Hardy and J. K. Hulm, *Phys. Rev.* **93**, 1004 (1954).
18. B. T. Matthias, *Phys. Rev.* **92**, 874 (1953).
19. O. I. Shulishova, *Izv. Akad. Nauk SSSR* **2**, 1434 (1966).
20. L. E. Toth, AFOSR 68-0265 (1968); Defense Doc. No. AD-671-944.
21. L. E. Toth, C. M. Yen, L. G. Rosner, and D. E. Anderson, *J. Phys. Chem. Solids* **27**, 1815 (1966).
22. C. M. Yen, L. E. Toth, Y. M. Shy, D. E. Anderson, and L. G. Rosner, *J. Appl. Phys.* **38**, 2268 (1967).
23. N. Pessall and J. K. Hulm, *Physics (Long Island City, N.Y.)* **2**, 311 (1966).
24. R. H. Willens, E. Buehler, and B. T. Matthias, *Phys. Rev.* **159**, 327 (1967).
25. L. E. Toth, E. Rudy, J. Johnston, and E. R. Parker, *J. Phys. Chem. Solids* **26**, 517 (1965).
26. J. Johnston, L. Toth, K. Kennedy, and E. R. Parker, *Solid State Commun.* **2**, 123 (1964).
27. J. Johnston, M.S. Thesis, University of California at Berkeley, 1964.
28. L. E. Toth, W. Jeitschko, and C. M. Yen, *Less-Common Metals* **10**, 29 (1966).
29. L. E. Toth, *J. Less-Common Metals* **13**, 129 (1967).
30. R. H. Willens and E. Buehler, *Appl. Phys. Lett.* **7**, 25 (1965).
31. B. T. Matthias, T. H. Geballe, and V. B. Compton, *Rev. Mod. Phys.* **35**, 1 (1963).
32. D. Gerstenberg and P. M. Hall, *J. Electrochem. Soc.* **111**, 936 (1964).
33. H. Bell, M.S. Thesis, University of Minnesota, 1966.
34. Y. M. Shy, M.S. Thesis, University of Minnesota, 1967.
35. H. Bell, Y. M. Shy, D. E. Anderson, and L. E. Toth, *J. Appl. Phys.* **39**, 2797 (1968).
36. T. Mitszuoka, T. Yamashita, T. Nakazawa, Y. Onodera, Y. Saito, and T. Anayama, *J. Appl. Phys.* **39**, 4788 (1968).
37. J. Zbasnik, L. E. Toth, Y. M. Shy, and E. Maxwell, *J. Appl. Phys.* **40**, 2147 (1969).
38. D. W. Deis, J. R. Gavaler, J. K. Hulm, and C. K. Jones, *J. Appl. Phys.* **40**, 2153 (1969).
39. H. J. Fink, A. C. Thorsen, E. Parker, V. F. Zackay, and L. Toth, *Phys. Rev.* **138**, 1170 (1965).
40. J. Piper, *Appl. Phys. Lett.* **6**, 183 (1965).
41. N. Pessall, C. K. Jones, H. A. Johansen, and J. K. Hulm, *Appl. Phys. Lett.* **7**, 38 (1965).
42. D. E. Anderson, L. E. Toth, L. G. Rosner, and C. M. Yen, *Appl. Phys. Lett.* **7**, 90 (1965).
43. K. Hechler, E. Saur, and H. Wizgall, *Z. Phys.* **205**, 400 (1967).
44. F. J. Darnell, P. E. Bierstedt, W. O. Forshey, and R. K. Waring, Jr., *Phys. Rev.* **A140**, 1581 (1965).
45. Y. Muto, N. Koto, T. Fukuroi, Y. Saito, T. Anayama, T. Mitszuoka, and Y. Onodera, *Appl. Phys. Lett.* **13**, 204 (1968).
46. E. Maxwell, B. B. Schwartz, and H. Wizgall, *Phys. Lett.* **A25**, 139 (1967).
47. M. Suenaga and K. M. Ralls, *J. Appl. Phys.* **37**, 4197 (1966).
48. R. R. Hake, *Appl. Phys. Lett.* **10**, 189 (1967).
49. K. Maki, *Phys. Rev.* **148**, 362 (1966).
50. N. R. Werthamer, E. Helfand, and P. C. Hohenberg, *Phys. Rev.* **147**, 295 (1966).
51. W. L. McMillan, *Phys. Rev.* **167**, 331 (1968).

52. T. H. Geballe, B. T. Matthias, J. P. Remeika, A. M. Clogston, V. B. Compton, J. P. Maita, and H. J. Williams, *Physics (Long Island City, N. Y.)* **2**, 293 (1966).
53. H. Bilz, *Z. Phys.* **153**, 338 (1958).
54. J. Piper, *in* "Compounds of Interest in Nuclear Reactor Technology" (J. T. Waber, P. Chiotti, and W. N. Miner, eds.), Inst. Metals Div., Spec. Rept. No. 13, p. 10. Edwards, Ann Arbor, Michigan, 1964.
55. L. E. Toth, V. F. Zackay, M. Wells, J. Olson, and E. R. Parker, *Acta Met.* **13**, 379 (1965).

8

Band Structure and Bonding in Carbides and Nitrides

I. Introduction

Bonding in transition metal carbides and nitrides is not completely understood. There are numerous contradictory theories of bonding for these phases, and available experimental evidence is insufficient to distinguish unambiguously between these theories. Nevertheless, certain features of bonding that have been clarified are useful in interpreting physical properties. Some of the properties that these theories should explain are listed below:

(1) The extremely high melting points of the fourth- and fifth-group carbides and to a lesser extent of the nitrides.

(2) The decreased thermodynamic stability of the sixth- and higher-group carbides and nitrides compared to the fourth- and fifth-group compounds. (The refractory compounds are in fact restricted to a rather limited portion of the periodic table.)

(3) The very high values of the elastic moduli and critical resolved shear stresses needed to induce slip.

(4) The apparent preference for the octahedral coordination polyhedron in binary carbides and nitrides and also in the ternary phases.

(5) The extensive range of solubilities in the phases; in some phases the

metal-to-nonmetal ratio varies by as much as 50%. Quite often, however, the solubility of the nonmetal in the structure of the parent transition metal is very limited.

(6) The ability of carbides and nitrides to form extensive solid solutions with each other, a fact which indicates a similarity in bonding.

(7) The observed superconducting, electrical and magnetic properties (see Chapters 6 and 7).

Bonding in transition-metal carbides and nitrides involves simultaneous contributions of metallic, covalent, and ionic bonding to the cohesive energy. Important bonds can be formed between both metal–metal pairs of atoms and metal–nonmetal pairs. Most theories have attempted to determine the relative importance of each type of contribution to bonding and the relative importance of metal–metal versus metal–nonmetal bonds. The theories, moreover, have been concerned about the direction of electron transfer, if any, between metal and nonmetal atoms.

Several theories have emphasized the metal–nonmetal bonds. These theories are based primarily upon crystallographic evidence of the strong preference for the octahedral coordination polyhedron by carbides and nitrides in both binary and ternary phases. The geometry of this polyhedron, together with the electron states available to the constituent atoms, is conducive to strong metal–nonmetal interactions.

Other theories have deemphasized the importance of metal–nonmetal bonding and emphasized instead metal–metal bonding, the interstitial nature of carbon or nitrogen, and the role of the nonmetal as a donator of electrons to the metal atoms. The interstitial solution theory has been used to explain the metallic conductivity of interstitial phases, the wide range of compositions over which these phases are stable, and the relatively small change in metal atom positions in the interstitial phases compared to these positions in the pure elements.

This difference in emphasis on the type of bonding depends upon the relative placement of the energies of the p-wave functions of the nonmetals. When the band arising principally from the p-wave functions is low in energy relative to the Fermi level, these p-like bands are occupied and strong bonding occurs between the sp electrons of the nonmetal and the d electrons of the metal. The strong nonmetal–metal bond lends support to the idea of strongly localized or covalent bonds; this concept of bonding is used to explain the hardness and brittleness of these compounds as well as the preference for the octahedral coordination polyhedron.

In the opposite approach, the p-wave functions of the nonmetal are placed high in energy relative to the Fermi level and are, therefore, not occupied. The d bands are filled instead; the bonding is primarily between metal atoms, with the nonmetal contributing electrons. The band structure resembles that found in the transition metals. This approach accounts for the high melting points of the compounds and the wide concentration ranges over which individual phases are stable. It also accounts for the decreased stability of these phases from group VI on.

II. Strong Metal–Nonmetal Bonding Theories

Rundle (*1*) was one of the first proponents of strong metal–nonmetal bonding. He observed that the distance between transition metal atoms (T—T distances) in octahedra is several percent greater than in the pure transition metal. He further noted that monocarbides and mononitrides nearly always have the *B*1 structure, regardless of the radius of the metal atom or the structure of the metal; the atomic packing of metal atoms has therefore changed to accommodate carbon or nitrogen in an octahedral interstitial site. Rundle suggested that the increased T—T distances greatly decrease the importance of T—T bonding and that the principal bonding mechanism is between the *sp* electrons of carbon and the d^2sp^3 electrons of the transition metal. Bonding between metal and nonmetal is accomplished by the use of the nonmetal *p* orbitals, or alternately by two *p* orbitals and two *sp* hybrid orbitals. In both cases, these orbitals are directed toward metal atoms at the corners of an octahedral coordination polyhedron (see Fig. 1).

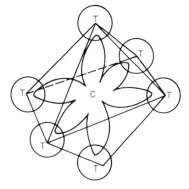

Fig. 1. Rundle's concept (*1*) of bonding in carbides emphasizes metal–nonmetal bonding and deemphasizes metal–metal bonds. Shown is the octahedral configuration with the bonding $sp–p^2$ configuration of the central carbon atom forming nonmetal-to-metal (XT) bonds with the transition atoms (T). [After Toth *et al.* (*2a*).]

The metal atom electrons are also used in these orbitals, and so electrons are, in effect, transferred from the metal to the nonmetal. No net ionic contribution occurs due to overlap of these orbitals on the metal atoms. Bonding is the *sp* type, with the formation of strong metal–nonmetal bonds.

Bilz (2) attempted to resolve the relative importance of metal–metal versus metal–nonmetal bonds. His LCAO calculation of the band structure considers nine interactions: $(ss\sigma)_T$, $(sd\sigma)_T$, $(dd\sigma)_T$, $(pp\sigma)_X$, $(pp\pi)_X$, $(dd\pi)_T$, $(pd\sigma)_{XT}$, $(ps\sigma)_{XT}$, and $(pd\pi)_{XT}$ where estimations for $(ss\sigma)_T$, $(dd\sigma)_T$, and $(dd\pi)_T$ are taken from values available for pure copper and nickel. Bilz maintains that his band structure is applicable to all NaCl structured refractory metals since he uses average values for the above two center integrals. Essentially, his band model consists of a narrow d-band (3.5 eV wide) superimposed over a broad s–p conduction band (see Fig. 2). He

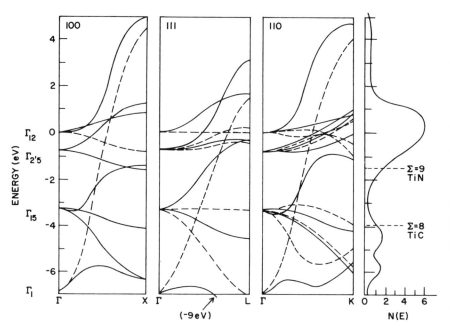

Fig. 2. Bliz's proposed band structure (2) and density of states histogram for TiC, TiN, and TiO with the NaCl crystal structure.

finds that the $(pd\sigma)_{XT}$ and $(ps\sigma)_{XT}$ two center integrals substantially lower the energies of the original p band of the interstitial atom so that electrons are transferred from the metal atoms to the nonmetal atoms ($2p$-bands). The d band is mainly empty for all refractory compounds from the fourth to sixth group. Bilz's calculations therefore support Rundle's hypothesis about the direction of electron transfer. Hardness, brittleness, and other properties result from the directional character of the XT interactions.

Bilz assumes in his calculations an average value for the one electron energies of the p electrons for C, N, and O and also average values of several transition elements for the one electron energies of the s and d electrons of the T atoms. Since he uses average values, Bilz claims that his band structure should apply to carbides, nitrides, and oxides. He states, furthermore, that the properties of the refractory interstitial compounds should be approximately determined by adjusting the height of the Fermi level according to the valence electron concentration (VEC). The VEC has been defined as α times the total number of valence electrons of the transition element plus β times the outer electrons of the interstitial. α and β are the real fraction of T and X sites occupied in $T_\alpha X_\beta$. Bilz further assumes that his band structure applies to nonstoichiometric NaCl-structured carbides and nitrides.

It is important to scrutinize these assumptions in more detail, since the Bilz band structure has been used to correlate many of the observed electrical and magnetic properties (Chapter 6). As also indicated in Chapter 6, many of the properties suggest that the stoichiometric carbides and nitrides can be treated as isoelectronic phases whose properties are accounted for by adjusting the height of the Fermi level according to the electron concentration.

The assumption that average values of one electron energies be used is not a good one. Table I lists the one electron energies for C, N, and O and

TABLE I

ONE ELECTRON ENERGIES OF T AND X ATOMS,
IN ELECTRON VOLTS [a]

	Ti	V	Zr	Nb
E_s	−7.1	−7.5	−7.4	−7.9
E_d	−9.1	−10.1	−8.3	
	C	N	O	
E_p	−10.7	−12.9	−15.9	

[a] Average values used by Bilz: $d_{0T} = 0$ (reference point); $s_{0T} = +1.5$ eV; $p_{0X} = -4.5$ eV.

the average value selected by Bilz, the average value is closest to that of O. Bilz's band structure probably better represents that of TiO than it does TiC. Furthermore, the large differences in the one electron energies suggest that band structures for the carbides, nitrides and oxides differ considerably. This criticism is consistent with much of the experimental data and several other theoretical models to be discussed.

Bilz did not have access to high-speed computers, and therefore, he estimated the density of states histogram. Using Bilz's values for the two-center integrals, we have recalculated the density of states histogram using 89 k vector-values appropriate to 1/48 of the Brillouin zone, which is equivalent to 2048 points in the full zone $(3)^1$ (see Fig. 3). The calculated

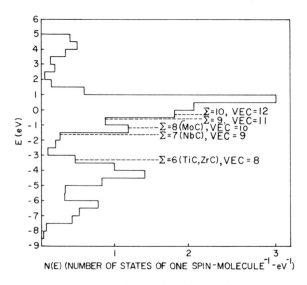

Fig. 3. A computer calculation of the density of states histogram in the Bliz's model shows that a minimum in the density of states occurs at valence electron concentrations between 8 and 9. The density of states for TiC, ZrC (VEC = 8) is higher, however, than that for VC, NbC (VEC = 9), a fact not indicated in the original Bliz estimation. The major difference between Bliz's estimated histogram and the one calculated here is the presence of a high energy band between 2 and 5 eV.

one differs from the Bilz estimate in the amount of detailed structure and in the presence of band between 2 and 5 eV. The present calculation also shows that the electron density of states at VEC = 8 (TiC, ZrC and HfC) is substantially higher than at VEC = 9 (NbC and TaC). Experimentally, however, low temperature heat capacity measurements give just the opposite results. Thus, the present calculations demonstrate that Bilz's choice for the two center integral values is not as consistent with the experimental data as his estimated histogram might indicate.

By extending the strong metal–nonmetal bonding concept, Denker (4, 5) has developed an interesting theory about the cause for appreciable lattice defects on both sublattices in transition metal carbides, nitrides and oxides.

---

[1] This calculation was performed by J. Zbasnik and L. E. Toth (unpublished data, 1968).

Using Bilz's band structure, he estimates that lattice vacancies reduce the effective valence electron concentration and lower the Fermi energy in TiO by about 1.75 eV. Since the energy to form a vacancy is estimated to be small, about 0.3 eV, vacancies are formed on both sublattices until the antibonding states are emptied. About 15% of all lattice sites are vacant in TiO. More detailed augmented plane wave (APW) calculations by Schoen (6) substantiate this interpretation about the reason for the high vacancy concentration. The same considerations could be applied to most carbides and nitrides which also tend to have vacancy concentrations even at stoichiometric compositions.

III. Strong Metal–Metal Bonding Theories

Several authors have proposed strong metal–metal bonding theories in carbides and nitrides. In particular the Russian school, lead by Samsonov (7, 8) and others, has been a strong advocate of the idea of the interstitial atoms acting as electron donors to strengthen the metal–metal bonds. Kiessling (9) also proposes an electron transfer in the opposite direction from that suggested by Rundle. Kiessling cites crystallographic, magnetic and thermodynamic evidence that the principal mechanism for bonding in these phases is between metal–metal atoms. The nonmetal donates electrons to the d bands of the transition metals.

Dempsey (10) has also adopted the view that carbides and nitrides are not covalently bonded materials but rather alloys similar to the transition metals from which they are derived. Dempsey bases his model on an examination of the melting points, T_m, of the carbides and nitrides and the transition metals. For the transition metals the maximum in T_m for any period in the periodic table occurs at about the sixth group (Cr, Mo, and W) (see Fig. 1, Chapter 1, p. 5). High melting points for this group are due to a filling of the bonding part of the d band, which contains roughly six electrons per atom (for a rough estimate of the band shape of transition metals see the variation of γ values with composition, Fig. 10, Chapter 6). The elements Cr, Mo, and W have the bonding portion of the d band nearly filled and hence exhibit a maximum in T_m. Elements belonging to groups before VI in the periodic table have not completely filled the bonding portion of the band, while elements after group VI have electrons in the antibonding portion of the band. In both of these cases, T_m of the elements is less than that for the sixth-group element. The only exception is vanadium, which melts at a slightly higher temperature than Cr.

According to Dempsey, transition metal carbides and nitrides have melting points comparable to those of transition elements. Only HfC, TaC, NbC, and Ta$_2$C have, in fact, melting points greater than that of tungsten. He argues, therefore, that carbides and nitrides are similar in bonding to transition metals. He observed, however, a shift in the position of the maximum in T_m with group number of the parent transition metal (see Fig, 1, Chapter 1). Fourth-group carbides and nitrides have melting points higher than those of the fifth and six group. Also, the chemical stability of carbides and nitrides decreases from the fourth group on (see Figs. 6 and 7, Chapter 4, p. 119). To explain these observations, Dempsey proposes that the p electrons from carbon and nitrogen are donated to the d bands and that these bands remain similar to their original configurations in pure transition metals. For a fourth-group carbide, there would be six electrons in the d band, filling the bonding portion of the band, while for higher-group carbides the antibonding portion of the d band is filled, which decreases both the melting point and the relative stability.

It was pointed out, however, in Chapter 6, that even if electrons are donated to these d bands, the γ values indicate an appreciably different band shape for carbides and nitrides than those found in the parent transition elements. Furthermore, soft X-ray emission spectra clearly indicates substantial differences between the band shape for carbides and that for the nitrides. These studies indicate that nitrogen is an electron acceptor.

Dempsey further supports his model on the basis of electrical resistivities, but as pointed out in Chapter 6, the resistivities of these phases are sensitive to the nonmetal-to-metal ratio. Without careful experimental studies of this effect for a sufficient number of phases, such relationships may be deceptive.

Costa and Conte (11) have also criticized the proposed direction of electron transfer from the metal to the nonmetal and Bilz's band model on the following basis:

> Bilz's model, which suppresses any binding between metal atoms, provides no explanation, in particular, as to why the transition metal carbides are much less stable from group VI onwards. The d band is practically empty in his model for TiC. It should be possible to observe considerable stabilization of the compounds for the metals of groups V, VI and VII similar to that found by filling the bonding half of the d band for the transition metals. In the same line of thought, one cannot understand why the existing structures always possess metal–metal distances approximately equal to those found in the metallic structure (12, p. 13).

In their theoretical model of the band structure, only the metal–metal interactions of the d_ϵ and d_γ type are considered[1]; metal–nonmetal inter-

---

[1] The d_ϵ states are d_{xy}, d_{yz}, and d_{zx} and the d_γ states are $d_{x^2y^2}$ and $d_{3z^2-r^2}$.

actions are neglected except for the effects of the nonmetal potential on the metal d states. The geometry of the d_ϵ and d_y functions is illustrated in Fig. 4. According to Costa and Conte, the d_ϵ states form a narrow band (~ 2.5 eV), while the d_y band is very broad (~ 10 eV) because of the effect of the nonmetal potential. The contribution of the interstitial atom to the potential in the interaction integrals between two d_y states is about four times greater than the contribution from the metal atoms. The interaction of d_y states may also be thought of as $(dp\sigma)_{XT}$ bonding because of the geo-

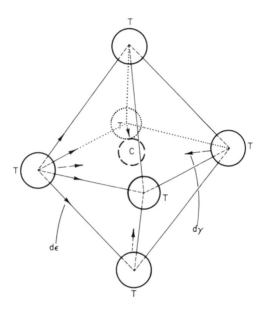

Fig. 4. Some of the directions of the d_y and d_ϵ states in an octahedron. The d_y states of the T atoms are directed toward the carbon atoms and interact strongly with the $2p$ states of carbon; the d_ϵ states are directed toward other T atoms and form strong T–T bonds. [After Toth *et al.* (2a).]

metry involved. The d_y states therefore form a broad bonding band. The d_ϵ states with a narrow band give shape to the entire d band. Electrons are transferred from the nonmetals to the d band. Since the Fermi level lies in the d band, electrical properties are explained in terms of the d band. Such properties as hardness and high melting points are explained, however, by the directional character of the d_y states and their interactions with the nonmetals.

Costa and Conte propose that C donates 1.5 electrons and N donates 2 electrons to the d bands of the transition metal. As stated in Chapter 6, this proposed direction for electron transfer in the carbides correlates well with the observed electrical, magnetic and optical properties. The properties do not, however, correlate well for the nitrides.

IV. Strong Metal–Metal and Metal–Nonmetal Bonds

The most recent theories of bonding for carbides and nitrides have emphasized the simultaneous contributions of metallic, covalent and ionic bonding to the cohesive energy. Both strong metal–metal and metal–nonmetal bonds are stressed. Even these theories, however, disagree with each other in considerable detail.

Lye (*13–15*) has considered bonding in TiC and has attempted to modify both the Bilz and Costa–Conte models by utilizing data from optical reflectivity experiments to adjust the relative electron-volt spacing between bands.

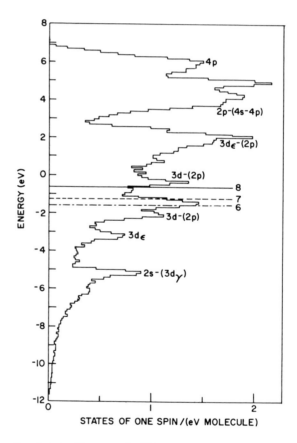

Fig. 5. Density-of-states histogram for TiC. Fermi levels are indicated for six, seven, and eight electrons; here eight electrons corresponds to stoichiometric TiC. [After Lye and Logothetis (*14*).]

He calculates a band structure by the LCAO method, a band model differing considerably from that of Bilz and one more general than that of Costa and Conte because of considerations of metal–nonmetal interactions. His calculated density-of-states curve is shown in Fig. 5.

The essential feature of Lye's model is that both metal–metal and metal–nonmetal interactions are important to the bonding and that the Fermi level in TiC lies in a band primarily of d character, with both the d_ϵ and d_γ states contributing to the shape of the band. The bonding states of the d band are depressed to energies lower than they are in pure hcp titanium metal. This depression increases the cohesive energy of the compounds and explains many properties such as high melting points. Because of the lowering of the bonding portions of the d bands, there are sufficient states in the d bands to accommodate the $3d$ and $4s$ electrons of titanium and also some of the $2p$ electrons of carbon. For TiC, Lye estimates the following electronic configurations for the atoms: for carbon $(2s)^2$ $(2p)^{3/4}$ and for titanium $(3d)^4(4s)^{3/4}(4p)^{1/2}$. About 1.25 electrons per atom are transferred from the carbon to the titanium atoms. The ionic charge is small, due to overlap of the metal functions on the carbon atoms.

Ern and Switendick (16) also emphasize metal–metal and metal–nonmetal bonding, but their band structure for TiC differs considerably from Lye's. They calculate the band structure of TiC and TiN by the augmented plane wave (APW) method. The advantage of this method is that no *a priori* assumptions are required about the degree of interaction between different states. The calculation can also be made self-consistent by comparing the assumed starting electronic configuration with the calculated one. The difficulty of the method for the carbides and nitrides is in estimating the potential on both metal and nonmetal atoms. That is, the method is self-consistent with the starting potential but this potential can change depending upon the assumed ionicity.

Figures 6 and 7 show the density of states histogram and the bonding character of each band as calculated by Ern and Switendick. For TiC, the $3d$ and $2p$ bands are strongly hybridized. For TiN the degree of hybridization is substantially reduced, and the $2p$ band is lower in energy than the Fermi level. These results show that one band model is inadequate for both the carbides and nitrides. It will be recalled that earlier models, e.g., Bilz's theory supposed that the band structure for carbides and nitrides could be suitably approximated by a rigid band model appropriate to both types of compounds.

The Ern and Switendick model predicts an electron transfer from the d bands to the $2p$-like states for both TiC and TiN and is therefore similar in this regard to the Bilz model. For TiC, however, their results do indicate strong metal–metal and metal–nonmetal bonding. Their model disagrees

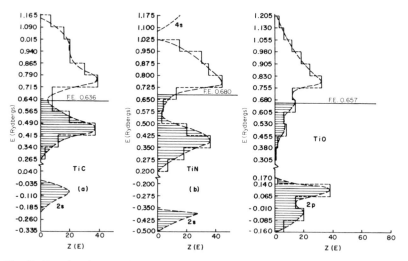

Fig. 6. Density of states histograms for TiC, TiN, and TiO. For TiC and TiO a hybridized $3d$–$2p$ band is split roughly into two halves with a broad minimum in the density of states separating each half. The Fermi level lies near this minimum. The $2s$ band for TiO is not shown. [After Ern and Switendick (16).]

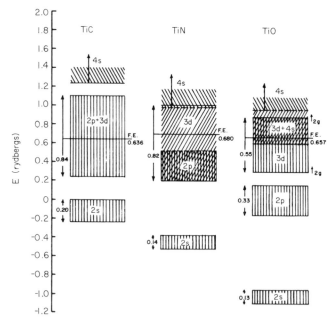

Fig. 7. Predominant character of each band in the Ern and Switendick band model. For TiC, the $3d$ and $2p$ states are strongly hybridized; for TiN and TiO the characteristic $2p$ band moves to lower relative energies. [After Ern and Switendick (16).]

with Lye's model in the relative placement of the $2p$-like band and the $4s$ band and therefore in the proposed direction of electron transfer.

For TiN the Ern and Switendick model indicates that some intermediate ionicity should have been assumed in the starting potential (N negative). For TiC any ionicity is small.

Conklin and Silversmith (17) attempted to resolve the controversy about the direction of charge transfer in TiC by performing APW calculations. They assumed different initial configurations for the potentials of the atoms. These corresponded to neutral atoms, to $Ti^{2+}C^{2-}$ and also to a nearly self-consistent potential. Their results, in general, confirm the Ern and Switendick (16) calculation and therefore indicate an electron transfer from Ti-like states to C-like states. There is, however, little net ionicity. The spatial distribution of the transferred electrons is such that the charge is nonlocally distributed.

V. Comparison with Observed Physical Properties

While there is not complete agreement on the band structure and bonding in carbides and nitrides, certain general features seem consistent with the observed physical properties. Although many features of the band structure are still very controversial areas of some agreement are listed below:

(1) Band structures for carbides and nitrides cannot be treated with a single rigid band model as originally proposed by Bilz. If the rigid band model is applicable, then the variation of electrical and magnetic properties caused by deviations from stoichiometry must be explained by the VEC parameter. Results of experimental investigations discussed in Chapters 6 and 7 show that the VEC parameter is inadequate to correlate these properties in the nonstoichiometric phases. These results also indicate that the degree of metal–nonmetal interactions and the direction of electron transfer between metal and nonmetal differs for carbides and nitrides.

(2) In the band structure, the d band is split into bonding and antibonding parts. Between these two parts, a low density of states exists and the Fermi level for most carbides and nitrides lies in this region. Whether or not this low density of states originates from sp or d electrons is not clear. This feature is found in most band models, except perhaps Lye's. Because his method of calculation is, however, least sensitive at the Fermi level, this difference does not contradict his theory (18). Most experiments on the electrical and magnetic properties clearly show that the density of states at the Fermi level for nearly all carbides and nitrides is low.

(3) Bonding in these compounds is a mixture of metal–metal and metal–nonmetal interactions. Both are important; it is difficult, if not impossible, to assign a relative importance to each.

(4) There is little ionicity in the carbides. While there is electron transfer from one type of quantum state to one whose character is due predominantly to another type of atom (as in the LCAO representation), the spatial distribution of these wave functions tends to minimize any net spatial charge transfer.

(5) For the nitrides, electrons are probably transferred from the metal d-like bands to the nitrogen $2p$ bands. The degree of ionicity is uncertain.

Areas of considerable controversy include the degree of hybridization and the relative energy placement of the $2p$ bands from carbon and the $4s$ band from Ti in TiC. Lye's band structure agrees with his optical reflectivity measurements; Ern and Switendick's band structure does not agree well with those measurements (18). Low-temperature heat-capacity measurements also indicate an electron transfer from C to the d bands in several non-stoichiometric carbide systems (19). Furthermore, the low-temperature heat capacities of nonstoichiometric carbides can be correlated with the Costa–Conte empirical parameters, assuming an electron transfer from C to the d bands of the metal (see Fig. 12, Chapter 6, p. 201).

In the beginning of this chapter, several properties were listed for which theories should offer explanations. Partial explanations can now be given in the same numerical order:

(1) The extremely high melting points of carbides are due to the splitting of bonding and antibonding parts of the d-like or p–d hybrid band. The melting points are high for the fourth and fifth group carbides because at these compositions the bonding parts of the band are just filled.

(2) The decreased thermodynamic stability of the sixth-group carbides and the fifth-group nitrides probably is the result of electrons beginning to fill the antibonding parts of the band. This decreased stability coincides with an increase in density-of-electron states.

(3) Their lack of ductility at low temperatures are probably due to covalent type metal–metal and metal–nonmetal hybrid bonds. The presence of the interstitial potential in the overlapping regions of the metal d functions increases the strength of the metal–metal interaction and may tend to localize it more. The observed slip planes in carbides $\{111\}$ are consistent with the hypothesis that the carbides are not primarily ionic.

(4) In addition to atomic size considerations, preference for the octahedral interstitial site in the crystallography stems from the strong nonmetal–metal bond formation (pd_γ hybrids) and the increased strength of the metal–metal bonds, due to d-like functions overlapping the interstitial potential.

(5) The extensive deviations from stoichiometry are consistent with the formation of strong metal–metal bonds. Limited evidence from mechanical property studies shows that the brittle-to-ductile transition temperature decreases with increasing deviations from stoichiometry in TiC (20) an indication that the metal–metal bonds become less localized and less strong as carbon is removed from the lattice.

(6) The ability of carbides and nitrides to form extensive solid solutions with each other is yet another indication that the bonding in these phases is similar.

(7) The observed superconducting, electrical, and magnetic properties are generally consistent with models that predict a low density of states at the Fermi level. The bonding and antibonding parts of the band structure of these phases are separated by a low density of states conduction band.

References

1. R. E. Rundle, *Acta Cryst.* **1**, 180 (1948).
2. H. Bilz, *Z. Phys.* **153**, 338 (1958).
2a. L. E. Toth, J. Zbasnik, Y. Sato, and W. Gardner, *in* "Anisotropy in Single-Crystal Refractory Compounds," Vol. I, p. 249 (F. W. Vahldiek and S. A. Mersol, eds.), Plenum Press, New York, 1968.
3. G. A. Burdick, *Phys. Rev.* **129**, 138 (1963).
4. S. P. Denker, *J. Phys. Chem. Solids* **25**, 1397 (1964).
5. S. P. Denker, *J. Less-Common Metals* **14**, 1 (1968).
6. J. M. Schoen, "Adaptation of the Augmented Plane Wave Method to Random, Three-Dimensional, Substitutional Alloys: Application to Titanium Monoxide," Ph.D. Thesis, Columbia University, 1969.
7. G. V. Samsonov and Ya. S. Umanskiy, "Tverdyye Soyedineniya Tugoplavkikh Metallov." State Sci.-Tech. Lit. Publ. House, Moscow, 1957; for English translation, see *NASA Tech. Trans.* **F-102** (1962).
8. Ya. S. Umanskiy, *Ann. Sect. Anal. Phys.—Chim., Inst. Chim. Gen.* **16**, No. 1, 128 (1943).
9. R. Kiessling, *Met. Rev.* **2**, 77 (1957).
10. E. Dempsey, *Phil. Mag.* [8] **8**, 285 (1963).
11. P. Costa and R. R. Conte, *in* "Compounds of Interest in Nuclear Reactor Technology" (J. T. Waber, P. Chiotti, and W. N. Miner, eds.), Inst. Metals Div., Spec. Rep. No. 13, p. 3. Edwards, Ann Arbor, Michigan, 1967.
12. P. Costa and R. R. Conte, *in* "Compounds of Interest in Nuclear Reactor Technology" (J. T. Waber, P. Chiotti, and W. N. Miner, eds.), Inst. Metals Div., Spec. Rep. No. 13, pp. 13–14. Edwards, Ann Arbor, Michigan, 1967.
13. R. G. Lye, RIAS Tech. Rep. 66–8 (1966).
14. R. G. Lye and E. M. Logothetis, *Phys. Rev.* **147**, 622 (1966).
15. R. G. Lye, *in* "Atomic and Electronic Structure of Metals," p. 99. Am. Soc. Metals, Metals Park, Ohio, 1967.

16. V. Ern and A. C. Switendick, *Phys. Rev.* **137**, A1927 (1965).
17. J. B. Conklin Jr. and D. J. Silversmith, *Int. J. Quantum Chem.* **2S**, 243 (1968).
18. R. G. Lye, private communication (1968).
19. L. E. Toth, M. Ishikawa, and Y. A. Chang, *Acta Met.* **16**, 1183 (1968).
20. G. E. Hollox, RIAS Tech. Rep. 68-10C (1968); *J. Mater. Sci. Eng.* **3**, 121 (1968–1969).

9

Postscript

I. Future Areas of Research and Critical Problems

Research areas on carbides and nitrides needing further development have been mentioned throughout this text. Several problems for future research are summarized below.

Diffusion studies in carbides and nitrides offer promising future investigations. Since present knowledge of this area is so limited and contradictory, the topic was not included in the main body of the text. It is important to understand the diffusion mechanism in these compounds and to obtain accurate values. Because the diffusion process is interrelated with many kinetic processes such as oxidation, diffusion studies have distinct practical value. Recently several investigators (1–4) have measured the self-diffusion of carbon by ^{14}C tracer methods, and they find relatively high activation energies (~ 110 kcal/mole). Other investigators (5) using layer-growth techniques find much lower activation energies (50–80 kcal/mole). This conflict should be resolved by further research. Attention should also focus upon isolating the effects of impurities (bound vacancy–impurity combinations) and stoichiometry on the results.

Preparation techniques for single crystals and thin films have only begun to be developed. High-purity single crystals of controlled compositions are needed for studies of mechanical behavior, diffusion, electrical and magnetic properties. High-purity thin films, probably best prepared by reactive

263

sputtering, offer new means of applying these materials. The characterization and physical properties of these films need to be explored further.

Better characterization of bulk samples is badly needed. Deviations from stoichiometry, crystal structure, lattice parameter, and impurity levels should be specified. Some present analytical techniques are unsatisfactory. Nitrogen analyses are difficult; most experimentalists perform these analyses in their own laboratories. Unfortunately, there are no standard samples to check their techniques. Perhaps the superconducting T_c might be used in some instances to help establish standards.

Significant progress is being made in defining interstitial ordering in the crystallography of carbides and nitrides. The effects of impurities on this ordering still need to be determined. Since ordering is apparently a very common phenomenon in these phases, its effects on physical properties should also be investigated.

The crystallography of thin films of carbides and nitrides is a relatively new field, but already several new structures have been found in the W–N system. Undoubtedly new crystal structures in thin-films will be found as this field expands.

Adequate theoretical treatments are needed to explain many observed thermodynamic regularities. The Schottky–Wagner model, as now formulated, is inadequate because it requires the heat of vacancy formation to be independent of composition. The theory should be reformulated to remove this restriction. In this regard, the theoretical efforts of Hoch (6), should be noted.

An experimental determination of the phonon spectrum of carbides and nitrides would be useful in understanding Debye temperatures, the superconducting interaction parameter, and thermal conductivity.

In the area of mechanical properties, practical considerations should dominate future research. Carbides have great intrinsic strength, but they cannot be utilized because of their brittleness. More research is needed on lowering the brittle-to-ductile transition temperature. More emphasis on nonstoichiometric phases may be promising.

In the area of superconductivity, emphasis should be on applications. Carbides and nitrides are excellent superconductors, but they are only beginning to be used in superconducting devices. Thin films offer the greatest promise, since they have excellent properties and can be used to overcome the intrinsic brittleness of these phases. Fast, sensitive bolometers using NbN have already been constructed (7) and miniature thin-film solenoids using NbN alloys are being developed (8). The main advantage of nitrides for this type of application is relative ease in preparing the films, compared to difficulties in preparing most other good superconducting materials.

In the areas of electrical and magnetic properties and bonding, more fundamental experimental studies on well-characterized samples are needed. Many measurements have been performed on poorly-characterized specimens; the effects of impurities, stoichiometry deviations, porosity, and vacancies are generally unknown. Single crystals would be most valuable for these measurements. More experiments on optical properties and on Lye's experimental approach (9) to band structure calculations are needed. The isoelectronic nature of these materials and crystal structure effects should be explored further. Stoichiometric high-purity single crystals, if possible to prepare, would be valuable for studying the Fermi surface and hence for formulating better theories of bonding.

References

1. S. Sarian and J. M. Criscione, *J. Appl. Phys.* **38**, 1794 (1967).
2. S. Sarian, *J. Appl. Phys.* **39**, 3305 (1968).
3. S. Sarian, *J. Appl. Phys.* **39**, 5036 (1968).
4. V. S. Eremeev and A. S. Panov, *Porosh. Met.* **7**, 65 (1967).
5. For a review of these values, see D. L. Harrod and L. R. Fleischer, *in* "Anistropy in Single-Crystal Refractory Compounds" (F. W. Vahldiek and S. A. Mersol, eds.), Vol. 1, p. 341. Plenum Press, New York, 1968.
6. See, e.g., M. Hoch, *in* "Phase Stability in Metals and Alloys" (P. S. Rudman, J. Stringer, and R. I. Jaffee, eds.), p. 419. McGraw-Hill, New York, 1967.
7. R. A. Smith, F. E. Jones, and R. P. Chasman, "The Detection and Measurement of Infrared Radiation." Oxford Univ. Press, London and New York, 1957.
8. L. E. Toth, University of Minnesota, private communication (1970).
9. R. G. Lye, *in* "Atomic and Electron Structure of Metals," p. 99. Am. Soc. Metals, Metals Park, Ohio, 1967.

Author Index

Numbers in parentheses are reference numbers and indicate that an author's work is referred to although his name is not cited in the text. Numbers in italics show the page on which the complete reference is listed.

A

Abrosimova, L. N., 95(63), *101*
Adelsberg, L. M., 74(21), 75, *100*, 165(39), *184*
Adelsköld, V., 61(41), 62(41), *68*
Agte, C., 87(42), *100*
Anayama, T., 20(36), *27*, 230(36), 236(36, 45), *245*
Anderko, K., 71(10), *99*
Anderson, D. E., 20(35), *27*, 86(33), *100*, 190(15), *213*, *216*, 225(21, 22), 226(21, 22), 230(35), 231(35), *232*, 233(21, 22), 234(21, 22), 236(21, 22, 35, 42), *245*
Anderson, O. L., *109*, 143, *183*
Andrews, K. W., 34, 35, *67*
Andrievskii, R. A., 165(42), *184*
Arkharov, V. I., 61(50), *68*
Armstrong, P. E., *109*, 146(14), *149*, *183*
Arnold, G. P., 82(30), 83(30), *100*
Aronsson, B., 29(1), *67*

B

Bangert, W. M., *113*
Bartlett, R. W., 145(12), *146*, *149*, *183*

Barton, J. W., 61(52), *68*
Baun, W. L., 208, *214*
Beattie, H. J., Jr., 35(10), *67*
Bell, H., 20(35, 37), *27*, *28*, 86(33, 34), *100*, 190(15), *213*, 230(33, 35), 231(33), 236(33, 35), *245*
Benesovsky, F., 3(1, 2), 4(11), 7(1, 11), 9(11), 11(1, 2), 12(1), 13(1, 2) *26*, *27*, 52(27, 28), *53*, 54(28), *54*, 55(27–29), 56(27–29), *56*, *57*, 58(28), 59(36, 37), *59*, 61(38), 63, *64*, 65(28, 57), *65*, 66(57), *66*, *67*, *68*, 70(5), 71(11), *82*, 83(5), *84*, 87(38), 88(38), 89, 90, 99(11, 12, 75), *99*, *100*, *101*, 175(50, 51), *184*, 186, *189*, 190, 196(21), 197(21), *213*, *214*
Berlincourt, T. G., *216*
Bernstein, B. T., *109*, 145(10), *146*, *183*
Bernstein, H., 121(36), *140*
Bierstedt, P. E., 236(44), 237(44), *236*
Bilz, H., 210(52), *214*, 243(53), 244, *246*, 250, *261*
Bittner, H., 186(10), 194(10), *195*, 196, 197, *213*, *214*
Blaugher, R. D., *216*
Blix, R., 96, *101*
Bloom, D. S., 79(25), *100*

267

Subject Index